T0257777

Handbook of Protein-Protein Interactions

Handbook of Protein-Protein Interactions

Edited by **Anton Torres**

New York

Published by Callisto Reference,
106 Park Avenue, Suite 200,
New York, NY 10016, USA
www.callistoreference.com

Handbook of Protein-Protein Interactions
Edited by Anton Torres

International Standard Book Number: 978-1-63239-411-8 (Hardback)

Printed in the United States of America.

Contents

Preface

This book provides an in-depth understanding of proteins and their interactions with each other. Proteins are vital in almost all biological procedures. The functions of proteins are synchronized through complex regulatory networks of transient protein-protein interactions (PPIs). To study PPIs, a broad range of methods have been developed over the past few decades. Several in vitro and in vivo attempts have been implemented to discover the mechanism of these pervasive interactions. Despite noteworthy developments in these investigational approaches, many issues exist such as false-positives/false-negatives, difficulty in obtaining crystal structures of proteins, etc. In order to overcome these challenges, various technical methods have been created, that are becoming more broadly used to examine PPIs. This book discusses various experimental approaches used to observe these interactions. It also looks at other important issues related to PPIs. The objective of this book is to bring together several researches accomplished by experts around the globe.

The researches compiled throughout the book are authentic and of high quality, combining several disciplines and from very diverse regions from around the world. Drawing on the contributions of many researchers from diverse countries, the book's objective is to provide the readers with the latest achievements in the area of research. This book will surely be a source of knowledge to all interested and researching the field.

In the end, I would like to express my deep sense of gratitude to all the authors for meeting the set deadlines in completing and submitting their research chapters. I would also like to thank the publisher for the support offered to us throughout the course of the book. Finally, I extend my sincere thanks to my family for being a constant source of inspiration and encouragement.

Editor

Part 1

Experimental Approaches

In Vivo Imaging of Protein-Protein Interactions

Hao Hong[1], Shreya Goel[2] and Weibo Cai[1]
[1]Departments of Radiology and Medical Physics,
University of Wisconsin - Madison, Madison, WI,
[2]Centre of Nanotechnology, Indian Institute of Technology, Roorkee,
[1]USA
[2]India

1. Introduction

Protein-protein interaction (PPI) plays a pivotal role in a wide variety of cellular events and physiological functions, such as enzymatic activity, signal transduction, immunological recognition, DNA repair/replication, among others (Valdar and Thornton, 2001). In addition, biological events that regulate proliferation, differentiation, and inflammation are also commonly mediated through PPI (Villalobos et al., 2007). Various techniques in molecular biology have been developed to understand the mechanism of these ubiquitous interactions, including qualitative methods such as yeast-two-hybrid screen (Fields and Song, 1989), immunoprecipitation (Williams, 2000), gel-filtration chromatography (Phizicky and Fields, 1995), etc. Meanwhile, quantitative biophysical methods have also been designed which include analytical ultracentrifugation (Hansen et al., 1994), calorimetry (Doyle, 1997), optical spectroscopy (Lakey and Raggett, 1998), etc. A decade ago, an assay for PPI based on β-galactosidase (gal) complementation was designed and successfully applied in cells (Wehrman et al., 2002).

Despite the success achieved by these techniques, none of them can be employed for interrogating PPI in living subjects due to several major limitations. First, traditional assays for measuring protein interactions require cell lysis, where the experimental conditions are inconsistent with the natural intracellular milieu. Second, these techniques may not be able to detect transient interactions that may have potent effects on cell signalling and intracellular processes. Lastly, the degree of false positives and false negatives vary from method to method, which significantly compromises the reproducibility and reliability of the data. With the tremendous expansion and evolution of the interdisciplinary field of molecular imaging over the last decade, many of these disadvantages have been or can be overcome.

Molecular imaging, "the visualization, characterization and measurement of biological processes at the molecular and cellular levels in humans and other living systems" (Mankoff, 2007), is an extremely powerful tool for imaging of PPI. The major molecular imaging modalities that have been applied for investigating PPI include bioluminescence, fluorescence, and positron emission tomography (PET) imaging. Quantitative and real-time molecular imaging of PPI can not only complement the already existing methodologies, which are mostly used in vitro or in cell culture, but also provide invaluable insights on PPI that were unavailable or impossible to investigate previously. For example, non-invasive imaging of PPI can dramatically accelerate the evaluation of new drugs in living subjects that promote or inhibit homodimeric/heterodimeric protein assembly (Massoud et al., 2007; Villalobos et al., 2007).

In this chapter, we will summarize the current status of in vivo imaging of PPI with various techniques, including fluorescence, bioluminescence, and PET imaging. A schematic summary of the most commonly used strategies for imaging of PPI are shown in **Figure 1**. To the best of our knowledge, there is no literature available on fluorescence imaging of PPI in animal models. However, since this is an indispensible component of imaging PPI in cell culture, herein we will give a few representative examples on fluorescence imaging of PPI to provide a complete overview of this dynamic research area.

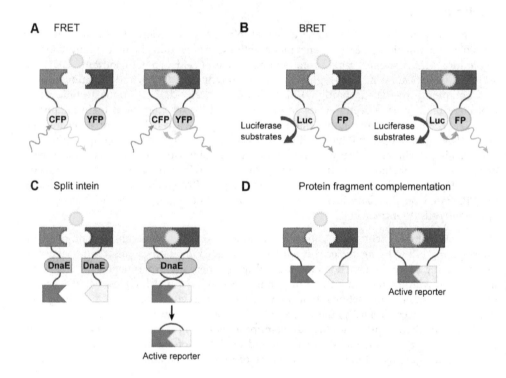

Fig. 1. Commonly used strategies for imaging of PPI. **A.** Fluorescence resonance energy transfer (FRET). **B.** Bioluminescence resonance energy transfer (BRET). **C.** Self-splicing split inteins (DnaE) can splice the two fragments of a reporter protein together into an intact and active reporter protein when they are brought within close proximity of each other.
D. Protein fragment complementation. Brown fragments are proteins of interest and the yellow star represents an inducer of PPI. Adapted from (Villalobos et al., 2007).

2. Fluorescence imaging of PPI

The (imaging) techniques used to detect or quantify PPI need to be sensitive within the concentration ranges at which proteins are present in cells or tissues, where sometimes fewer than 10^4 protein molecules may be present. Furthermore, these techniques should be capable of identifying interactions of specific proteins against a background of more than 30,000 other proteins within a living cell. As a technology that has had an impact on almost all areas of biology, fluorescent imaging can meet these criteria under certain scenarios and has been widely used for imaging of PPI in vitro.

Fluorescence spectroscopy and fluorescence imaging have been demonstrated to be versatile tools for imaging of PPI. Fluorescent proteins (FPs), specifically variants of the green FP (GFP), are among the most frequently used for imaging of PPI (Giepmans et al., 2006; van Roessel and Brand, 2002). In a typical fluorescence process, an electron in the fluorophore within the FP absorbs photons from suitable excitation light (in the UV or visible range), which raises the energy level of the electron to an excited state. During this short excitation period, some of the energy is dissipated through molecular collisions or transferred to a proximal molecule, and the remaining energy is emitted as a photon to relax the electron back to the ground state (van Roessel and Brand, 2002). Since the energy is lower for the emission photon than the excitation photon, the emission wavelength is longer than the excitation wavelength which can be readily separated by applying a filter of specific wavelength range.

Fluorescence imaging of PPI in cell culture has the potential to provide information on the cellular and sub-cellular distribution of FPs with sub-second time resolution. Fluorescence microscopy techniques, primarily including fluorescence resonance energy transfer (FRET) and fluorescence correlation spectroscopy (FCS), are commonly used to quantify the activity, interaction, and dynamics of protein molecules within living cells (Yan and Marriott, 2003). Many protein interactions are transient, or energetically weak, thereby precluding their identification and analysis through traditional biochemical methods such as co-immunoprecipitation. In this regard, the genetically encodable FPs (GFP, yellow FP [YFP], cyan FP [CFP], red PP [RFP], etc.) and their associated overlapping fluorescence spectra have granted researchers the ability to monitor weak interactions in live cells using FRET.

2.1 Imaging of PPI with FRET

FRET requires the measurement of the relative intensity of the emission signal from a pair of fluorophores (Tsien, 2009). The underlying physics is attributed to a quantum mechanical effect between a given pair of fluorophores (i.e. a fluorescent donor and an acceptor) where, upon excitation of the donor, energy is transferred from the donor to the acceptor in a non-radiative manner by means of dipole-dipole coupling (Jares-Erijman and Jovin, 2003). Upon energy transfer, donor fluorescence is quenched and acceptor fluorescence is increased (sensitized), resulting in a decrease in donor excitation lifetime. The FRET efficiency is the quantum yield of the energy transfer transition, i.e. the fraction of energy transfer event occurring per donor excitation event, which is dependent upon several factors including the distance between the donor and the acceptor, the spectral overlap of the donor emission spectrum and the acceptor absorption spectrum, as well as the relative orientation of the donor emission dipole moment and the acceptor absorption dipole moment.

Since FRET is critically dependent upon molecular proximity, it has been described as a molecular ruler. FRET typically operates in a range of 1-10 nm, a distance that is relevant for most molecules engaged in complex formation or conformational changes. FRET from CFP to YFP is a commonly used strategy for monitoring protein interactions or conformational changes of individual proteins. For example, FRET-based assays involving CFP and YFP were designed and employed to monitor receptor interactions on endothelial cells in one report (Seegar and Barton, 2010). However, one disadvantage of FP-based FRET is that protein functions may be perturbed by fusion of FPs since they are quite large in size. In one study, G protein-coupled receptor (GPCR) activation in living cells was used as a model system to compare YFP with a small fluorescent agent (FlAsH), which was targeted to a short tetracysteine sequence (Hoffmann et al., 2005). It was found that FRET from CFP to FlAsH reports GPCR activation in living cells without disturbing receptor function, which is more advantageous than the use of YFP as the FRET acceptor.

FRET has also been employed to visualize the interaction between two FPs, enhanced GFP (EGFP) and mCherry (Albertazzi et al., 2009). One- and two-photon fluorescence lifetime imaging microscopy (FLIM) were used to determine the FRET efficiency values. It was found that this FP pair can be used for effective and quantitative FRET imaging of PPI. Since FLIM can produce images based on the differences in the exponential decay rate of the fluorescence signal from different fluorophores, advances in FRET and FLIM have enabled studies of PPI at the microscopic level. FLIM provides a promising and robust method of detecting molecular interactions via FRET by monitoring the variation of donor fluorescence lifetime, which is insensitive to many factors that can influence the conventional intensity-based measurements, such as fluorophore concentration, photobleaching, spectral bleed-through, donor-acceptor stoichiometry, light path length etc. (Pelet et al., 2006; Zhong et al., 2007). The fact that FRET can deplete the excited state population of the donor and cause a reduction in both its fluorescence intensity and lifetime makes this technique well suited for studies in intact cells.

Interrogating PPI deep inside living tissues requires precise fluorescence lifetime measurements to derive the FRET between two tagged fluorescent markers. In a recent study, FLIM was used in combination with a clinically licensed remote endoscopic cellular resolution imaging modality to map PPI in live cells embedded in a 3D matrix, which served as a model of a diseased organ structure in a patient (Fruhwirth et al., 2010). This strategy allowed accurate measurement of fluorescence lifetime changes on the order of 100 ps, which not only demonstrated the feasibility of studying PPI by FRET in cultured living cells within 3D matrices, but also provided potential instrumentation for other FRET-based assays.

The FRET/FLIM technique can also provide invaluable information for the mechanistic study of PPI in different types of diseases. In one study which investigated the mechanism of metastasis induction by the S100A4 protein, interactions of S100A4 with C-terminal recombinant fragment of non-muscle myosin heavy chain in living HeLa cells were mapped using confocal microscopy, FLIM, and time-correlated single-photon counting (Zhang et al., 2005). The findings indicated that not only there is direct interaction between S100A4 and its target in live mammalian cells, but also that such an interaction contributes to metastasis induction, thus shedding new light onto the mechanism of cancer metastasis. In another report, FRET/FLIM enabled the study of the interaction between hypoxia-inducible factor-1α (HIF-1α) and HIF-2α with the aryl hydrocarbon receptor nuclear translocator in a hypoxia model, which provided new information about specific gene expression controlled by PPI in hypoxia (Konietzny et al., 2009). FRET/FLIM has also been employed to image dynamic PPI in neurons (**Figure 2**), which enhanced the understanding of nervous system development and function (Timm et al., 2011). Protein kinases of the microtubule affinity regulating kinase (MARK)/Par-1 family play important roles in the establishment of cellular polarity, cell cycle control, and intracellular signal transduction. Disturbance of their function is linked to cancer and various brain diseases. In this recent study, transfected Teal FP (TFP) and YFP were used as FRET donor and acceptor pairs in neurons and imaged by FLIM, which revealed that MARK was particularly active in the axons and growth cones of differentiating neurons (Timm et al., 2011).

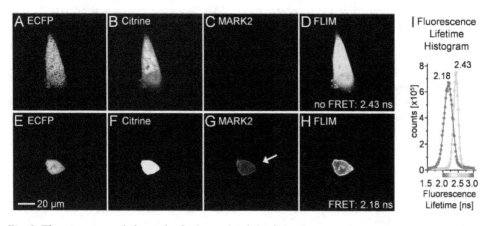

Fig. 2. The upper panel shows both channels of the fluorescence intensity image (**A, B**) of a cell transfected with a construct composed of ECFP (i.e. enhanced CFP) linked to Citrine (i.e. a stable variant of YFP), which does not exhibit FRET in the absence of fluorescently labeled MARK2 (i.e. the inducer of FRET) as indicated by a lack of fluorescence signal in **C**. The pseudo-colored FLIM image is shown in **D**, which has a long fluorescence lifetime of 2.43 ns. FRET between the two FPs (**E, F**) occurs when MARK2 is present, as indicated by the fluorescence signal in **G**. The short fluorescence lifetime of 2.18 ns is shown as red in **H** (high FRET). The graph **I** displays the averaged histograms of cells showing FRET (red dots) or no FRET (green dots) and gaussian fits of the data. Reprinted with permission from (Timm et al., 2011).

Not limited to the imaging of PPI, FRET can also be employed for imaging protein-DNA interactions, such as through the use of near-infrared fluorescent oligodeoxyribonucleotide reporters that can sense transcription factor NF-κB p50 protein binding (Zhang et al., 2008). Recently, a similar approach using hairpin-based FRET probes for the detection of human recombinant NF-κB p50/p65 heterodimer binding to DNA was reported (Metelev et al., 2011). Both of these studies demonstrated that FRET-based technique can give signal changes that are simple to interpret and stoichiometrically correct for detecting transcription factor-DNA interactions.

2.2 Imaging of PPI with FCS

Different from FRET, FCS detects the diffusion rate of single molecules which can give insights regarding whether a protein is part of a larger complex or not (Elson, 2004; Haustein and Schwille, 2007). Based on the analysis of intensity fluctuation of one or a few labeled protein conjugates at nanomolar concentration in a femtoliter volume, which depends on several factors such as the number of fluorescent species in the excitation volume, the diffusion constant of the conjugate, etc., FCS has been used to study PPI, protein-lipid/ligand-receptor interactions, dimerization of membrane receptors and proteins involved in the downstream signalling, DNA dynamics, among others (Elson, 2004; Haustein and Schwille, 2007). The high sensitivity and the possibility to monitor these dynamic interactions makes FCS a powerful tool to study signal transduction in cellular or even tissue environment at physiologically relevant conditions (Hink et al., 2002).

FCS is relatively insensitive to molecular mass. Therefore, species with similar molecular weight cannot be differentiated. Dual color fluorescence cross-correlation spectroscopy (FCCS) measures interactions by cross-correlating two or more fluorescent channels (one channel for each molecule/protein of interest), which can distinguish interactions and dynamics of biomolecules more sensitively than FCS, particularly when the mass change in the reaction/interaction is small. However, the inherent drawback of FCCS is that it suffers from non-ideal confocal volume overlap and spectral cross-talk which severely limits its applications. Fluorescence lifetime correlation/cross-correlation spectroscopy has the potential to resolve this issue, as demonstrated in a recent study (Chen and Irudayaraj, 2010). Interaction of a fluorescently-labeled antagonist antibody with the epidermal growth factor receptor (EGFR)-GFP construct in live HEK293 cells were monitored by both fluorescence lifetime cross-correlation measurements and FLIM, which not only opens up new opportunities in studying PPI in solutions and in live cells but also provides new biological insights in understanding how an antagonist influences EGFR through live cell imaging and quantification.

The field of plant sciences has also benefited from these techniques mentioned above. For example, FRET/FLIM was used to investigate CDC48A, a member of the AAA ATPases (i.e. ATPases associated with diverse cellular activities) family which has various functions in cell division, membrane fusion, as well as proteasome- and ER-associated degradation of proteins (Aker et al., 2007). With the use of FCS, it was shown that CDC48A hexamers are part of even larger complexes.

2.3 Imaging of PPI with other fluorescence techniques

Besides FRET/FLIM and FCS, enzyme complementation was also adopted for fluorescence imaging of PPI a decade ago (Spotts et al., 2002). A reporter technology based on the differential induction of β-lactamase (Bla) enzymatic activity was developed to function as a sensor for the interaction state of two target proteins within single neurons. Bla was split into two separate, complementary protein fragments which can be brought together by phosphorylation-dependent association of the kinase inducible domain of the cyclic adenosine monophosphate (cAMP) response element binding (CREB) protein and the KIX domain of the CREB binding protein (Spotts et al., 2002). Using an intracellular substrate whose fluorescence spectrum changes upon hydrolysis by Bla, time-lapse ratiometric imaging measurements were achieved after association of CREB and CREB binding protein, which permits direct imaging of PPI in single cells with high signal discrimination.

To investigate the conformational changes of proteins in living cells when external force is applied, a genetically encoded fluorescent sensor was constructed and tested in a myosin-actin model system using the proximity imaging (PRIM) technique, which detects spectral changes of two GFP molecules that are in direct contact (Iwai and Uyeda, 2008). The developed PRIM-based strain sensor module (PriSSM), consisted of the tandem fusion of a normal and circularly permuted GFP, was inserted between two motor domains of dictyostelium myosin II to study the effect of strain. It was suggested that this technology may provide a general approach for studying force-induced protein conformational changes in cells.

2.4 A brief summary of fluorescence imaging of PPI

The FRET/FLIM technique can be used as a versatile tool to characterize the spatial distribution of various proteins and detect/quantify PPI in a living cell, which can measure intermolecular FRET through quite sophisticated mathematical algorithms. However, no in vivo fluorescence imaging of PPI has been reported so far since these techniques (in particular FP-based) cannot be readily used for in vivo imaging applications due to several major limitations.

First, FRET-based techniques require the use of incident light to activate the donor protein. Given that the excitation wavelength is typically in the green range, little excitation light will travel through tissue since most tissues have strong light absorption/attenuation below a wavelength of 600 nm (Frangioni, 2003). Therefore these techniques are intrinsically not suitable for non-invasive imaging studies in live animals. Second, there is strong auto-fluorescence signal from animal tissue which confounds the interpretation of the imaging data. Third, the sensitivity of fluorescence imaging is not very high. Fourth, the relative molar ratios of the FRET donor/acceptor pair are not always 1:1, which can cause significant problems in calibration, detection, and quantification, especially when the situation is exacerbated in vivo when compared to cell-based studies. Lastly, there is significant photobleaching when the FPs are exposed to excitation light for a prolonged period.

3. Bioluminescence imaging (BLI) of PPI

Because of very low background signal and high sensitivity, BLI can be a more suitable technique for in vivo imaging of PPI than fluorescence imaging. The fact that no additional excitation light will be needed in BLI is highly advantageous for reducing the background signal. Two major strategies have been adopted for BLI of PPI: bioluminescence resonance energy transfer (BRET) and enzyme complementation.

3.1 Imaging of PPI with BRET

BRET displays similar characteristics as FRET except the donor is a bioluminescent protein, typically a luciferase, which requires the presence of small molecule substrates but not incident light. Similar to FRET, BRET is also a quantum process in which energy is transferred over a distance, usually < 10 nm, from the donor (e.g. luciferase) to a FP

(Villalobos et al., 2007). However, BRET offers many distinct advantages over FRET because of its higher quantum yield and better detection sensitivity.

As a popular technique for studying PPI in live cells, BRET is particularly suitable for real-time monitoring of such interactions. For example, many cellular signal transduction can be visualized by this technique, such as agonist-induced GPCR/β-arrestin interaction (Pfleger et al., 2006), calcium sensing receptor homodimer formation (Jensen et al., 2002), β2-adrenergic receptor dimerization (Angers et al., 2000), interaction of circadian clock proteins (Xu et al., 1999), etc. Since the potential for studying the modulation of such interactions by agonists, antagonists, inhibitors, dominant negative mutants, and co-expressed accessory proteins is tremendous, high-throughput BRET-based screening system is an ever-expanding area of interest for the pharmaceutical industry. However, imaging PPI with BRET in animal models is very challenging and only a few successful examples are available in the literature (Massoud et al., 2007; Villalobos et al., 2007).

In one early study, a cooled charge-coupled device (CCD) camera-based spectral imaging strategy enabled simultaneous visualization and quantitation of BRET signal from live cells and cells implanted in living mice, where renilla luciferase (RLuc) and its substrate were used as an energy donor and a mutant GFP was used as the acceptor (De and Gambhir, 2005). As a proof-of-principle, the donor and acceptor proteins were fused to FKBP12 and FRB respectively, which are known to interact only in the presence of the small molecule mediator rapamycin (Banaszynski et al., 2005; Choi et al., 1996). Mammalian cells expressing these fusion constructs were imaged using a cooled-CCD camera either directly from culture dishes or after implanting them into mice, where the specific BRET signal was determined by comparing the emission photon yields in the presence and absence of rapamycin. Such CCD camera-based imaging of BRET signal is very appealing since it can seamlessly bridge the gap between in vitro and in vivo studies, thus validating BRET as a powerful tool for interrogating and detecting PPI directly at limited depths in living mice.

Subsequently, a highly photon-efficient and self-illuminating fusion protein, which combines a mutant RFP (mOrange) and a mutant RLuc (RLuc8), was constructed to improve the BRET efficiency/signal (De et al., 2009). This new BRET fusion protein, termed as "BRET3", exhibited several-fold improvement in light intensity when compared with the previous BRET fusion proteins. In addition, BRET3 also exhibits red-shifted light output, which can allow for deeper tissue imaging in small animals. At single cell level, the BRET3 construct (which contains FKBP12 and FRB) was demonstrated to only exhibit BRET signal in the presence of rapamycin. With increased photon intensity, red-shifted light output and good spectral resolution (approximately 85 nm), it was suggested that BRET3-based assays will allow imaging of PPI using a single assay that is directly scalable from living cells to small animals.

Recently, further improvement on the BRET3 construct was reported, which was termed "BRET6" (Dragulescu-Andrasi et al., 2011). Red light-emitting BRET-based reporter systems were developed to allow for assaying PPI both in cell culture and in deep tissues of small animals (**Figure 3**). These BRET systems consist of the newly developed RLuc variants (RLuc8 and RLuc8.6, which serve as BRET donors) and two RFPs (TagRFP and TurboFP635, which serve as BRET acceptors). In addition to the native coelenterazine substrate for RLuc, a synthetic derivative (coelenterazine-v) was also used which further red-shifted the emission maxima of RLuc by 35 nm. Ratiometric imaging of PPI in the presence of rapamycin-induced FKBP12-FRB association was demonstrated in both cultured cells and small animal tumor models.

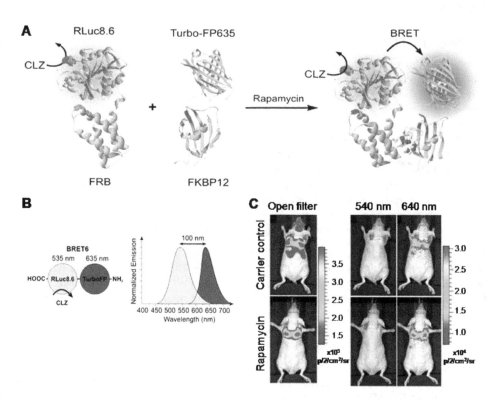

Fig. 3. Imaging of PPI with BRET6. **A.** Schematic illustration of the BRET6 construct for monitoring rapamycin-induced FRB-FKBP12 association. **B.** Schematic representation of the BRET6 fusion construct, the emission spectrum of the RLuc mutant, and the absorption spectrum of the acceptor protein. CLZ denotes coelenterazine (a substrate for RLuc). **C.** Bioluminescence images of cells stably expressing the BRET6 construct, accumulated in the lungs of nude mice after intravenous injection. Mice were also injected with both rapamycin (or control carrier which does not contain rapamycin) and CLZ before imaging. Adapted from (Dragulescu-Andrasi et al., 2011).

Currently, the number of BRET probes reported for the imaging of PPI is significantly lower when compared to FRET-based approaches. Much future work needs to be devoted to BRET-based imaging of PPI. The strategy of combining a fluorescent and a bioluminescent reporter to generate self-illuminated reporter proteins is advantageous to overcome the common problems associated with in vivo fluorescent imaging of PPI. As a genetically encodable approach for ratiometric imaging of PPI in cells and living subjects, light attenuation by tissue is the major challenge for ratiometric analysis of PPI with a BRET system. Since light attenuation varies with the wavelength of the emitted photons and the tissue depth, red-shifted luciferases and FPs are clearly preferred choices. Meanwhile, consistency of the BRET ratio in different mice should also be monitored carefully to ensure sufficient spatial control to retain the ratiometric characteristics of a BRET sensor.

3.2 Imaging of PPI with complementation of split enzyme

Enzyme complementation assay depends on the division of a reporter enzyme (e.g. luciferase) into two separate inactive components that can regain function upon association (Massoud et al., 2007). When the two enzyme fragments are each fused to two interacting proteins, the enzyme can be reactivated upon PPI. For in vivo BLI applications, the split firefly luciferase (fLuc) with small overlapping sequences is a suitable choice because it consistently yields strong signal and excellent inducible complementation by a variety of PPIs. The reaction kinetics and ease of delivery of the substrate, D-luciferin, also allows for facile application of this technique in BLI assays. Besides fLuc, RLuc has also been investigated for BLI of PPI. However, although the split RLuc system functions quite efficiently, one major limitation of RLuc-based assay is its substrate, coelenterazine, which exhibits poor reaction profile for long-term kinetic experiments. In addition, the hydrophobicity of the molecule also makes it difficult to use for in vivo applications.

The first report on non-invasive BLI of PPI in living subjects based on a split luciferase was achieved a decade ago (Paulmurugan et al., 2002). In this study, split fLuc was designed and constructed for both intein-mediated reconstitution and complementation, where the two fLuc fragments could be brought together by the strong interaction between two proteins, MyoD and Id, both of which are members of the helix-loop-helix family of nuclear proteins. As a demonstration of the proof-of-principle, cells transiently transfected with the split reporter gene construct were used for imaging MyoD-Id interactions, both in cell culture and in cells implanted into living mice.

In a subsequent study, the split fLuc strategy was employed for imaging of PPI in hypoxia (Choi et al., 2008). HIF-1α is well known to regulate the activation of genes that promote malignant progression (Koh et al., 2010). HIF-1α is hydroxylated on prolines 402 and 564 under normoxia, which is targeted for ubiquitin-mediated degradation by interacting with the von Hippel-Lindau protein complex (pVHL). To study the interaction between HIF-1α and pVHL, the split fLuc-based system was used where HIF-1α and pVHL were fused to the amino-terminal and carboxy-terminal fragments of fLuc, respectively. Hydroxylation-dependent interaction between HIF-1α and pVHL led to complementation of the two fLuc fragments, resulting in bioluminescence in vitro and in vivo. Complementation-based bioluminescence was diminished when mutant pVHL with decreased binding affinity for HIF-1α was used. This strategy represents a useful approach for studying PPI involved in the regulation of protein degradation. In another study, split fLuc was also used for investigating epidermal growth factor (EGF)-induced Ras/Raf-1 interaction in mammalian cells (Kanno et al., 2006).

Similar strategy has been adopted to develop an inducible split RLuc-based bioluminescence assay for quantitative measurement of real time PPI in mammalian cells (Paulmurugan and Gambhir, 2003). In a follow-up study, the split RLuc construct was used to evaluate drug-modulated PPI in a cancer model in living mice (**Figure 4**)

(Paulmurugan et al., 2004). The heterodimerization of FRB and FKBP12, mediated by rapamycin, was also utilized in this study. The concentration of rapamycin needed for efficient dimerization, as well as the amount of ascomycin (a competitive binder of rapamycin) required for dimerization inhibition, were investigated. These studies demonstrated that such split reporter-based strategies can be used to efficiently screen small molecule drugs that modulate PPI, and further evaluate the effect of the drugs in living animals.

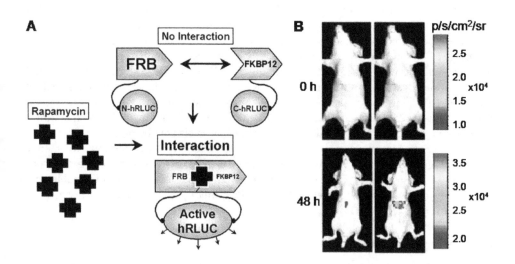

Fig. 4. In vivo imaging of drug-modulated PPI. **A.** Schematic diagram of rapamycin-mediated complementation of the two fragments of synthetic renilla luciferase (hRLUC). **B.** Non-invasive imaging of PPI in living mice, intravenously injected with human 293T embryonic kidney cancer cells that were transiently co-transfected with both split constructs. Mice not receiving rapamycin (left) showed only background signal, whereas the animals receiving repeated injections of rapamycin emitted higher signals originating from the 293T cells in the liver (right). Adapted from (Paulmurugan et al., 2004).

Homodimeric PPI, potent regulators of cellular functions and particularly challenging to study in vivo, can also be visualized by the split RLuc strategy. Split RLuc complementation-based bioluminescence assay was used to study the homodimerization of herpes simplex virus type 1 thymidine kinase (HSV1-TK) in mammalian cells and in living mice (Massoud et al., 2004). Homodimerization of HSV1-TK chimeras containing the N-terminal or C-terminal fragments of RLuc in the upstream and downstream positions, respectively, was visualized and quantified. A mutant of HSV1-TK was used to confirm the specificity of the RLuc complementation signal from HSV1-TK homodimerization. This generalizable assay to screen for molecules that promote or disrupt ubiquitous homodimeric PPI can not only serve as an invaluable tool to understand the biological signaling networks, but will also be useful in drug discovery/validation in live animal disease models. In a cell-based study, the split RLuc strategy was shown to be useful beyond the visualization and confirmation of the existence of PPI. It also helped in identifying the critical amino acid residues involved in a specific PPI (Jiang et al., 2010).

Besides fLuc and RLuc complementation, split click beetle luciferase has been used to study the interaction between GPCR and β-arrestin (Misawa et al., 2010), whereas split Gaussia luciferase has been employed to image the interaction between calmodulin and other proteins (Kim et al., 2009). However, neither of these split luciferases has been demonstrated for in vivo visualization of PPI. Other split enzymes have also been explored for the imaging of PPI, such as the use of split β-gal for BLI of GPCR interactions in vivo (von Degenfeld et al., 2007). Currently, there is a paucity of sensitive and specific methods for quantitative comparison of the pharmacological properties of GPCRs in physiological and/or pathological settings in live animals. In this study, low affinity and reversible β-gal complementation was developed to quantify GPCR activation via interaction with β-arrestin, which enabled real time BLI of GPCR activity in live animals with high sensitivity and specificity (von Degenfeld et al., 2007). Imaging was achieved by using a recently developed luminescent β-gal substrate, which is a caged luciferin molecule that can be recognized by fLuc to generate light only after it has been cleaved by β-gal (Wehrman et al., 2006). Following implantation of the cells into mice, it was possible to monitor pharmacological GPCR activation and inhibition in their physiological context by non-invasive BLI, suggesting that this technology may have unique advantages to enable novel applications in the functional investigation of GPCR modulation in biological research and drug discovery.

4. PET imaging of PPI

Typically, PPI represents a low-level biological event and is therefore very challenging to detect, locate, and image in intact living subjects. When compared with BLI and fluorescence imaging, PET possesses very high sensitivity, while being quantitative and tomographic (Massoud and Gambhir, 2003). In addition, it is one of the few non-invasive imaging techniques that can be applied in humans for non-invasive monitoring of reporter gene expression (Kang and Chung, 2008). Although PET has enormous potential in imaging complex biological events such as PPI, to date only one example of PET imaging of PPI has been reported (Massoud et al., 2010).

The PET reported gene HSV1-TK was molecularly engineered and cleaved between Thr265 and Ala266, where the fragments were used in a protein-fragment complementation assay to quantify as well as to non-invasively image PPI in mammalian cells and living mice (Massoud et al., 2010). It was found that a point mutation (V119C) could be introduced to markedly enhance the HSV1-TK complementation modulated by several different PPIs such as the rapamycin-mediated FKBP12- FRB, HIF-1α-pVHL, etc. In vivo PET imaging of the FKBP12-FRB interaction modulated through rapamycin was successfully achieved (**Figure 5**). Future applications of this unique split HSV1-TK strategy are potentially far reaching, including accurate monitoring of immune and stem cell therapies, as well as allowing for fully quantitative and tomographic PET localization of PPI in preclinical small and large animal models of various diseases.

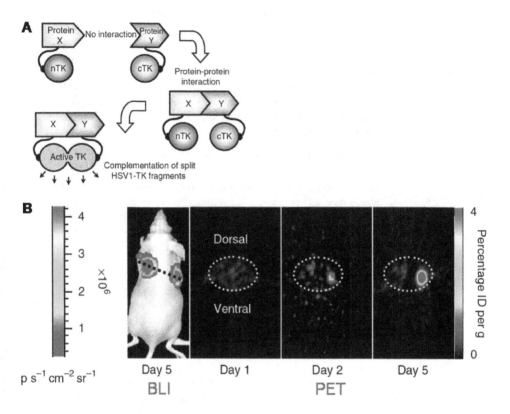

Fig. 5. Non-invasive PET imaging of PPI. **A.** Schematic diagram showing the use of split HSV1-TK to monitor the hypothetical X-Y heterodimeric PPI. **B.** Transaxial PET images of a mouse implanted subcutaneously with mock-transfected 293T cells (left) and 293T cells stably expressing both split constructs of HSV1-TK each fused to FRB and FKBP12 respectively (right). The serial images at different days were acquired after injection of the PET reporter probe for HSV1-TK (i.e. ^{18}F-FHBG). A BLI image of the mouse is also shown to delineate the two tumors. Adapted from (Massoud et al., 2010).

5. Conclusion

The interactions of specific cellular proteins form the basis of a wide variety of biological processes, including many signal transduction and hormone activation pathways involved in maintaining important biological functions. Accurate measurement of PPI can significantly help in deciphering the genetic and proteomic code. The tremendous complexity of cellular events requires assays that can measure different types of PPIs using an array of different methods. Molecular imaging, an extremely powerful tool to study molecular events in living subjects, can provide invaluable information and insight in elucidating the process of various PPIs.

To date, the major molecular imaging modalities used for visualization of PPI include fluorescence imaging (not suitable for in vivo studies), BLI, and PET imaging. All these techniques require extensive efforts in protein engineering due to the complex and challenging nature of imaging PPI in living cells and animals. Particularly for split reporter-based strategies, intensive efforts are needed to obtain better functioning split reporters that exhibit efficient PPI-induced complementation but not self-complementation. At the same time, sufficiently high reporter activity needs to be maintained upon PPI-induced complementation. For in vivo imaging of PPI, PET serves as a better choice over BLI and fluorescence due to its superb sensitivity, excellent tissue penetration, high quantification accuracy, and potential for clinical translation.

Future work on the imaging of PPI may include the design of second-generation complementation reporters with improved signal-to-noise ratios, inducibility, and red-shifted spectral properties for more wide spread use in vivo. The ideal reporter for imaging of PPI should not only serve as an "on/off" signal, but also give a graduated and quantitative response with minimal background signal and excellent induced signal output. Lastly, since no single imaging modality is perfect, combination of different imaging techniques to study the same PPI may provide complementary information.

6. References

Aker, J.; Hesselink, R.; Engel, R.; Karlova, R.; Borst, J.W.; Visser, A.J. & de Vries, S.C. (2007). In vivo hexamerization and characterization of the Arabidopsis AAA ATPase CDC48A complex using forster resonance energy transfer-fluorescence lifetime imaging microscopy and fluorescence correlation spectroscopy. *Plant Physiol*, 145, 339-350.

Albertazzi, L.; Arosio, D.; Marchetti, L.; Ricci, F. & Beltram, F. (2009). Quantitative FRET analysis with the EGFP-mCherry fluorescent protein pair. *Photochem Photobiol*, 85, 287-297.

Angers, S.; Salahpour, A.; Joly, E.; Hilairet, S.; Chelsky, D.; Dennis, M. & Bouvier, M. (2000). Detection of beta 2-adrenergic receptor dimerization in living cells using bioluminescence resonance energy transfer (BRET). *Proc Natl Acad Sci USA*, 97, 3684-3689.

Banaszynski, L.A.; Liu, C.W. & Wandless, T.J. (2005). Characterization of the FKBP.rapamycin.FRB ternary complex. *J Am Chem Soc*, 127, 4715-4721.

Chen, J. & Irudayaraj, J. (2010). Fluorescence lifetime cross correlation spectroscopy resolves EGFR and antagonist interaction in live cells. *Anal Chem*, 82, 6415-6421.

Choi, C.Y.; Chan, D.A.; Paulmurugan, R.; Sutphin, P.D.; Le, Q.T.; Koong, A.C.; Zundel, W.; Gambhir, S.S. & Giaccia, A.J. (2008). Molecular imaging of hypoxia-inducible factor 1 alpha and von Hippel-Lindau interaction in mice. *Mol Imaging*, 7, 139-146.

Choi, J.; Chen, J.; Schreiber, S.L. & Clardy, J. (1996). Structure of the FKBP12-rapamycin complex interacting with the binding domain of human FRAP. *Science*, 273, 239-242.

De, A. & Gambhir, S.S. (2005). Noninvasive imaging of protein-protein interactions from live cells and living subjects using bioluminescence resonance energy transfer. *FASEB J*, 19, 2017-2019.

De, A.; Ray, P.; Loening, A.M. & Gambhir, S.S. (2009). BRET3: a red-shifted bioluminescence resonance energy transfer (BRET)-based integrated platform for imaging protein-protein interactions from single live cells and living animals. *FASEB J*, 23, 2702-2709.

Doyle, M.L. (1997). Characterization of binding interactions by isothermal titration calorimetry. *Curr Opin Biotechnol*, 8, 31-35.

Dragulescu-Andrasi, A.; Chan, C.T.; De, A.; Massoud, T.F. & Gambhir, S.S. (2011). Bioluminescence resonance energy transfer (BRET) imaging of protein-protein interactions within deep tissues of living subjects. *Proc Natl Acad Sci USA*, 108, 12060-12065.

Elson, E.L. (2004). Quick tour of fluorescence correlation spectroscopy from its inception. *J Biomed Opt*, 9, 857-864.

Fields, S. & Song, O. (1989). A novel genetic system to detect protein-protein interactions. *Nature*, 340, 245-246.

Frangioni, J.V. (2003). *In vivo* near-infrared fluorescence imaging. *Curr Opin Chem Biol*, 7, 626-634.

Fruhwirth, G.O.; Ameer-Beg, S.; Cook, R.; Watson, T.; Ng, T. & Festy, F. (2010). Fluorescence lifetime endoscopy using TCSPC for the measurement of FRET in live cells. *Opt Express*, 18, 11148-11158.

Giepmans, B.N.; Adams, S.R.; Ellisman, M.H. & Tsien, R.Y. (2006). The fluorescent toolbox for assessing protein location and function. *Science*, 312, 217-224.

Hansen, J.C.; Lebowitz, J. & Demeler, B. (1994). Analytical ultracentrifugation of complex macromolecular systems. *Biochemistry*, 33, 13155-13163.

Haustein, E. & Schwille, P. (2007). Fluorescence correlation spectroscopy: novel variations of an established technique. *Annu Rev Biophys Biomol Struct*, 36, 151-169.

Hink, M.A.; Bisselin, T. & Visser, A.J. (2002). Imaging protein-protein interactions in living cells. *Plant Mol Biol*, 50, 871-883.

Hoffmann, C.; Gaietta, G.; Bunemann, M.; Adams, S.R.; Oberdorff-Maass, S.; Behr, B.; Vilardaga, J.P.; Tsien, R.Y.; Ellisman, M.H. & Lohse, M.J. (2005). A FlAsH-based FRET approach to determine G protein-coupled receptor activation in living cells. *Nat Methods*, 2, 171-176.

Iwai, S. & Uyeda, T.Q. (2008). Visualizing myosin-actin interaction with a genetically-encoded fluorescent strain sensor. *Proc Natl Acad Sci USA*, 105, 16882-16887.

Jares-Erijman, E.A. & Jovin, T.M. (2003). FRET imaging. *Nat Biotechnol*, 21, 1387-1395.

Jensen, A.A.; Hansen, J.L.; Sheikh, S.P. & Brauner-Osborne, H. (2002). Probing intermolecular protein-protein interactions in the calcium-sensing receptor homodimer using bioluminescence resonance energy transfer (BRET). *Eur J Biochem*, 269, 5076-5087.

Jiang, Y.; Bernard, D.; Yu, Y.; Xie, Y.; Zhang, T.; Li, Y.; Burnett, J.P.; Fu, X.; Wang, S. & Sun, D. (2010). Split Renilla luciferase protein fragment-assisted complementation (SRL-PFAC) to characterize Hsp90-Cdc37 complex and identify critical residues in protein/protein interactions. *J Biol Chem*, 285, 21023-21036.

Kang, J.H. & Chung, J.K. (2008). Molecular-genetic imaging based on reporter gene expression. *J Nucl Med*, 49 Suppl 2, 164S-179S.

Kanno, A.; Ozawa, T. & Umezawa, Y. (2006). Intein-mediated reporter gene assay for detecting protein-protein interactions in living mammalian cells. *Anal Chem*, 78, 556-560.

Kim, S.B.; Sato, M. & Tao, H. (2009). Split Gaussia luciferase-based bioluminescence template for tracing protein dynamics in living cells. *Anal Chem*, 81, 67-74.

Koh, M.Y.; Spivak-Kroizman, T.R. & Powis, G. (2010). HIF-1alpha and cancer therapy. *Recent Results Cancer Res*, 180, 15-34.

Konietzny, R.; Konig, A.; Wotzlaw, C.; Bernadini, A.; Berchner-Pfannschmidt, U. & Fandrey, J. (2009). Molecular imaging: into in vivo interaction of HIF-1alpha and HIF-2alpha with ARNT. *Ann N Y Acad Sci*, 1177, 74-81.

Lakey, J.H. & Raggett, E.M. (1998). Measuring protein-protein interactions. *Curr Opin Struct Biol*, 8, 119-123.

Mankoff, D.A. (2007). A definition of molecular imaging. *J Nucl Med*, 48, 18N, 21N.

Massoud, T.F. & Gambhir, S.S. (2003). Molecular imaging in living subjects: seeing fundamental biological processes in a new light. *Genes Dev*, 17, 545-580.

Massoud, T.F.; Paulmurugan, R.; De, A.; Ray, P. & Gambhir, S.S. (2007). Reporter gene imaging of protein-protein interactions in living subjects. *Curr Opin Biotechnol*, 18, 31-37.

Massoud, T.F.; Paulmurugan, R. & Gambhir, S.S. (2004). Molecular imaging of homodimeric protein-protein interactions in living subjects. *FASEB J*, 18, 1105-1107.

Massoud, T.F.; Paulmurugan, R. & Gambhir, S.S. (2010). A molecularly engineered split reporter for imaging protein-protein interactions with positron emission tomography. *Nat Med*, 16, 921-926.

Metelev, V.; Zhang, S.; Tabatadze, D. & Bogdanov, A. (2011). Hairpin-like fluorescent probe for imaging of NF-kappaB transcription factor activity. *Bioconjug Chem*, 22, 759-765.

Misawa, N.; Kafi, A.K.; Hattori, M.; Miura, K.; Masuda, K. & Ozawa, T. (2010). Rapid and high-sensitivity cell-based assays of protein-protein interactions using split click beetle luciferase complementation: an approach to the study of G-protein-coupled receptors. *Anal Chem*, 82, 2552-2560.

Paulmurugan, R. & Gambhir, S.S. (2003). Monitoring protein-protein interactions using split synthetic renilla luciferase protein-fragment-assisted complementation. *Anal Chem*, 75, 1584-1589.

Paulmurugan, R.; Massoud, T.F.; Huang, J. & Gambhir, S.S. (2004). Molecular imaging of drug-modulated protein-protein interactions in living subjects. *Cancer Res*, 64, 2113-2119.

Paulmurugan, R.; Umezawa, Y. & Gambhir, S.S. (2002). Noninvasive imaging of protein-protein interactions in living subjects by using reporter protein complementation and reconstitution strategies. *Proc Natl Acad Sci USA*, 99, 15608-15613.

Pelet, S.; Previte, M.J. & So, P.T. (2006). Comparing the quantification of Forster resonance energy transfer measurement accuracies based on intensity, spectral, and lifetime imaging. *J Biomed Opt*, 11, 34017.

Pfleger, K.D.; Dromey, J.R.; Dalrymple, M.B.; Lim, E.M.; Thomas, W.G. & Eidne, K.A. (2006). Extended bioluminescence resonance energy transfer (eBRET) for monitoring prolonged protein-protein interactions in live cells. *Cell Signal*, 18, 1664-1670.

Phizicky, E.M. & Fields, S. (1995). Protein-protein interactions: methods for detection and analysis. *Microbiol Rev*, 59, 94-123.

Seegar, T. & Barton, W. (2010). Imaging protein-protein interactions in vivo. *J Vis Exp*, pii: 2149.

Spotts, J.M.; Dolmetsch, R.E. & Greenberg, M.E. (2002). Time-lapse imaging of a dynamic phosphorylation-dependent protein-protein interaction in mammalian cells. *Proc Natl Acad Sci USA*, 99, 15142-15147.

Timm, T.; von Kries, J.P.; Li, X.; Zempel, H.; Mandelkow, E. & Mandelkow, E.M. (2011). Microtubule affinity regulating kinase (MARK) activity in living neurons examined by a genetically encoded FRET/FLIM based biosensor: Inhibitors with therapeutic potential. *J Biol Chem*, Epub ahead of print.

Tsien, R.Y. (2009). Indicators based on fluorescence resonance energy transfer (FRET). *Cold Spring Harb Protoc*, 2009, pdb top57.

Valdar, W.S. & Thornton, J.M. (2001). Protein-protein interfaces: analysis of amino acid conservation in homodimers. *Proteins*, 42, 108-124.

van Roessel, P. & Brand, A.H. (2002). Imaging into the future: visualizing gene expression and protein interactions with fluorescent proteins. *Nat Cell Biol*, 4, E15-20.

Villalobos, V.; Naik, S. & Piwnica-Worms, D. (2007). Current state of imaging protein-protein interactions in vivo with genetically encoded reporters. *Annu Rev Biomed Eng*, 9, 321-349.

von Degenfeld, G.; Wehrman, T.S.; Hammer, M.M. & Blau, H.M. (2007). A universal technology for monitoring G-protein-coupled receptor activation in vitro and noninvasively in live animals. *FASEB J*, 21, 3819-3826.

Wehrman, T.; Kleaveland, B.; Her, J.H.; Balint, R.F. & Blau, H.M. (2002). Protein-protein interactions monitored in mammalian cells via complementation of beta -lactamase enzyme fragments. *Proc Natl Acad Sci USA*, 99, 3469-3474.

Wehrman, T.S.; von Degenfeld, G.; Krutzik, P.O.; Nolan, G.P. & Blau, H.M. (2006). Luminescent imaging of beta-galactosidase activity in living subjects using sequential reporter-enzyme luminescence. *Nat Methods*, 3, 295-301.

Williams, N.E. (2000). Immunoprecipitation procedures. *Methods Cell Biol*, 62, 449-453.

Xu, Y.; Piston, D.W. & Johnson, C.H. (1999). A bioluminescence resonance energy transfer (BRET) system: application to interacting circadian clock proteins. *Proc Natl Acad Sci USA*, 96, 151-156.

Yan, Y. & Marriott, G. (2003). Analysis of protein interactions using fluorescence technologies. *Curr Opin Chem Biol*, 7, 635-640.

Zhang, S.; Metelev, V.; Tabatadze, D.; Zamecnik, P.C. & Bogdanov, A., Jr. (2008). Fluorescence resonance energy transfer in near-infrared fluorescent oligonucleotide probes for detecting protein-DNA interactions. *Proc Natl Acad Sci USA*, 105, 4156-4161.

Zhang, S.; Wang, G.; Fernig, D.G.; Rudland, P.S.; Webb, S.E.; Barraclough, R. & Martin-Fernandez, M. (2005). Interaction of metastasis-inducing S100A4 protein in vivo by fluorescence lifetime imaging microscopy. *Eur Biophys J*, 34, 19-27.

Zhong, W.; Wu, M.; Chang, C.W.; Merrick, K.A.; Merajver, S.D. & Mycek, M.A. (2007). Picosecond-resolution fluorescence lifetime imaging microscopy: a useful tool for sensing molecular interactions in vivo via FRET. *Opt Express*, 15, 18220-18235.

2

Protein-Protein Interactions in Salt Solutions

Jifeng Zhang

Department of Analytical and Formulation Sciences, Amgen Inc.,
Thousand Oaks, California
USA

1. Introduction

Protein-protein interactions drive many biophysical processes of proteins in solutions, such as aggregation, solubility, and phase transitions including crystallization, gelation, and amorphous precipitation. Many of these processes are of significant research interest because of their practical importance. In the biopharmaceutical industry, it is crucial to prevent therapeutic proteins from aggregation during the manufacturing process and storage in order to maintain safety and efficacy (1). In addition, protein crystallization and precipitation are used for industrialized recombinant protein purification process (2). In the field of structure biology, it is still a daunting task to produce diffractive quality protein crystals for determining protein 3-D structures because there is lack of clear understanding of the mechanisms for protein crystallization (3). Furthermore, studying protein-protein interactions could shed light on the mechanism of protein condensation (or phase transition) diseases, such as cataract and sickle cell disease (4). Finally, protein-protein interactions may play essential roles in many human neurodegenerative diseases attributed to protein aggregation, such as Parkinson and Alzheimer diseases (5).

In solutions, salts are ubiquitously used to control pH, ionic strength and osomlality in scientific research and industry applications. It is important to understand how salts modulate protein-protein interactions so that solution behavior, such as protein crystallization, precipitation, and solution stability, can be controlled and manipulated. However, the exact interaction mechanisms between salt ions and proteins are poorly understood (6, 7). As a consequence, modulations on protein-protein interactions by salt ions and their implications for protein solution behavior cannot be completely rationalized. The challenges rise because of (i) the sheer complexity of physical and chemical properties for both salt ions and proteins and (ii) the wide range of salt concentrations, which can be varied up to 1000 fold from millimolar to molar. It cannot be emphasized better than how Kunz and Neueder mentioned in their book with regards to salt solutions: "In total, it is still a fact that over the last decades, it was still easier to fly to the moon than to describe the free energy of even the simplest salt solutions beyond a concentration of 0.1 M or so"(6). Proteins probably belong to the most complex colloidal system in terms of variations in surface charge, surface chemistry, and size. Specifically, a protein could be net positive-charged, neutral, or negative-charged at pH conditions below, near, and above its pI (Isoelectric point), respectively. Additionally, protein surfaces are heterogeneously composed of positive and negative charged, polar and nonpolar amino acid residues. Finally, the size of proteins in the range of 1-5 nm (estimated by the minimal radius of a sphere containing a given mass) would significantly impact the surface charge density(8).

Intermolecular interactions between protein molecules can have different origins, such as electrostatic, hydrophobic, van der waals, and hydrogen bonding (9). It is difficult to pinpoint the exact relative contributions from each type of interaction to the (overall) protein-protein interactions. In this review, I focus on explaining the modulations of electrostatic protein-protein interactions by the simple salt ions (shown in Figure 1) through their specific interactions (or binding) from both cation and anion with protein surface at salt concentrations below 0.5- 1 M. In addition, the complete picture of salt ion's effects on the intermolecular interactions may be better understood by considering the following biophysical properties of proteins and salt ions: (i) the net charge, surface charge density and hydrophobicity of a protein; (ii) hydration, size, polarizability and valency of salt ions. The discussion is based on the recent experimental results reported in literature and findings from Amgen using the following experimental techniques, such as protein solubility measurement, phase transition temperature of $T_{critical}$ (critical temperature) or T_{cloud} (cloud temperature) for liquid-liquid phase separation and small angle X-ray scattering (SAXS) (10-13). It has been demonstrated that there is a strong correlation between protein solubility and protein-protein interactions: protein solubility decreases when the protein-protein interactions become less repulsive or more attractive (for a protein for which its solubility increases with temperature)(12, 13). Also it is generally accepted that for a protein solution with an upper consolute point, an increase in phase transition temperature, as a result of change in the solution condition, indicates that protein-protein interactions become less repulsive or more attractive.

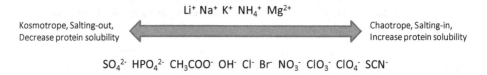

$$Li^+ \ Na^+ \ K^+ \ NH_4^+ \ Mg^{2+}$$

Kosmotrope, Salting-out,   Chaotrope, Salting-in,
Decrease protein solubility   Increase protein solubility

$$SO_4^{2-} \ HPO_4^{2-} \ CH_3COO^- \ OH^- \ Cl^- \ Br^- \ NO_3^- \ ClO_3^- \ ClO_4^- \ SCN^-$$

Fig. 1. Hofmeister series adapted from (14).

2. Historical background

2.1 Direct and reverse Hofmeister series

The most important experimental work on protein-protein interactions in salt solutions can be traced back more than 100 years ago when Franz Hofmeister and his coworkers studied salt effects at high salt concentrations on protein precipitation of hen egg white proteins whose main component is ovalbumin (pI=4.6). At that time, he hypothesized that the protein precipitating (salting-out) capability for the salts was dependent on their ion hydration properties (6). Later on, an empirical ranking for both cations and anions in their effectiveness, as shown in Figure 1, for precipitating proteins was named as (direct) Hofmeister series (14). Typically, the anions' effects are more dramatic than cation (14). In 1989, a surprising and complete reverse Hofmeister series was discovered by Ries-Kautt and

Ducruix in solubility measurement of lysozyme in salt solutions at pH below its pI where the protein was net positively charged (15).

2.2 Protein-protein interactions for a net charge neutral protein in salt solutions

A protein is net-charge neutral at its pI with the equal numbers of positive and negative charges. This is the most distinctive difference between proteins and the peptides with neutral side chains/small nonpolar molecules, for which extensive and detailed solubility experiments were conducted in salt solutions (16-19). However, there is lack of systematic protein solubility studies in salt solutions near their pIs. It is generally accepted that near the pI an increase in protein solubility (salting-in) is expected when salts are initially added and then a decrease occurs at high salt concentrations (salting-out by kosmotropic salts)(20). Although the mechanism of protein-protein interactions near its pI remains to be determined, it can be inferred from the observation above that the protein-protein interactions may initially become less attractive and then more attractive with increasing salt concentrations.

2.3 Protein-protein interactions for a net positive-charged protein in salt solutions

Lysozyme is a small globular protein with a Molecular Weight (MW) of 14.4 kilo-Dalton (kD) with a high pI value of ~11(12). Despite the fact that the experiments can mostly be conducted at pH conditions below its pI, lysozyme was frequently used as a model protein for studying both protein-protein interactions and protein-salt ion interactions in salt solutions probably due to its availability and easy crystallization propensity. Numerous experiments revealed very complex relationships between intermolecular interactions and salt concentration, salt type and pH; different theories were put into place to interpretate the trends (12, 21, 22).

In monovalent salt solutions under 1.0 M, the intermolecular interactions for lysozyme generally became monotonically more attractive as the salt concentration increased at pH conditions far below its pI(12, 21). These findings are consistent with the no salting-in event, i. e. protein solubility decrease, for lysozyme by NaCl in a pH range from 3 to 9 under the salt concentration up to 1.2 M (23). Acting as counter-ions to the net positively-charged lysozyme and following the reverse Hofmeister series, these monovalent anions imposed profound effects on the intermolecular interactions. But at pH 9.4 closer to pI, a nonmonotonic transition was discovered for SCN- where the intermolecular interactions initially became more attractive and then less attractive when the phase transition temperature was measured (22). For γD-crystallins, a 20-kD protein, the same reverse Hofmeister series for anions was observed at pH 4.5 below its pI of ~7.0 by using SAXS (13).

Despite the dominant effect of the counter-ions (or anions), the co-ions (or cations) can still significantly perturb the protein-protein interactions. Specifically, comparing the effect by different cation in the salt solutions with the same anion, the intermolecular interactions for positive-charged lysozyme were less attractive and even perturbed nonmonotonically by the strongly hydrated divalent cation (Mg^{2+} and Ca^{2+}) , in comparison to the monotonic effect by the monovalent cations of Na^+ and K^+(12, 21). These findings are consistent with the findings from lysozyme solubility measurement in the multivalent cation salt solutions (12, 24).

2.4 Protein-protein interactions for a net negative-charged protein in salt solutions

Recently, many experiments were conducted to study protein-protein interactions for a net negatively-charged protein in salt solutions where a cation-dominant effect was expected. But the experimental findings were not straight-forward to interpret. Using SAXS and neutron scattering for studying protein-protein interactions of ovalbumin (MW=45 kD) in NaCl and YCl$_3$ solutions at pH conditions above its pI of 5.2, it was found that NaCl was ineffective in screening the electrostatic repulsive interactions between the proteins while YCl$_3$ not only suppressed the electrostatic repulsive interactions initially but also raised the repulsive interactions at higher concentrations (25). The ineffectiveness of Na$^+$ salts to screen the electrostatic repulsion was also confirmed for α-crystallins, a 800-kDa protein, at pH conditions above its pI of 4.5 by using SAXS (13). Similar behaviour was observed for BSA at pH conditions above its pI of 4.6 (26). Interestingly, Petsev et al found that NaAcetate was effective at screening the electrostatic repulsions (protein-protein interactions become more attractive) and then rendered the intermolecular interactions more repulsive for negatively-charged Apoferrtin (MW=450kD) (27).

2.5 Protein-protein interactions for an antibody at different pH conditions

Protein-protein interactions in salt solutions for an antibody with an experimentally determined pI of 7.2 were systematically explored through the measurements of protein solubility and phase transition temperature of $T_{critical}$ in liquid-liquid phase separation (11). The advantage of using this antibody is that the intermolecular interactions can be systemically assessed for the positive-charged and neutral for the same protein, allowing comprehensive experimental investigations of how salts modulate intermolecular interactions. Also, the antibody (MW=147 kD) is a much larger protein than lysozyme, which provides an opportunity for evaluating the surface charge density as a variable in protein-protein interactions(10). These approaches could help us understand how salt ions interact with proteins of different size.

At pH 7.1 close to its pI of 7.2, antibody solubility measurement revealed a general salting-in effect by all the anions as shown in Figure 2. More importantly, the specific anion

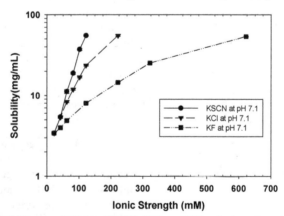

Fig. 2. Antibody solubility at pH 7.1 in KSCN, KCl and KF solutions [reprint with permission from ref (11)].

effect was observed in which SCN⁻ was the most effective at raising the antibody solubility, following the direct Hofmeister series. These observations are consistent with the ranking of these anions for disrupting the attractive intermolecular interactions as revealed by the results of $T_{critical}$ measurement (10).

At pH 5.3 below its pI, nonmonotonic behavior where protein solubility decreased and then increased with salt concentrations (in Figure 3) was observed for all the salts studied, suggesting that intermolecular interactions became less repulsive and then more. In addition, the effectiveness of the anions for reducing the protein solubility followed the reverse Hofmeister series, in which SCN⁻ was the most effective at reducing the antibody solubility. Then strikingly, the effectiveness for the anion to increase the protein solubility reverted back to the direct Hofmeister series as the salt concentration further increased. The above nonmonotonic transitions are in agreement with the protein-protein interactions pattern revealed by the measurement of $T_{critical}$ for liquid-liquid phase separation in the same salt solutions (10).

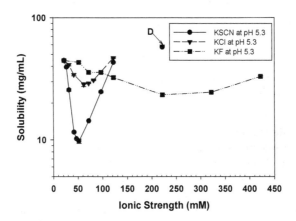

Fig. 3. Antibody solubility at pH 5.3 in in KSCN, KCl and KF solutions [reprint with permission from ref (11)].

It should be interesting to further study how salts affect the antibody solubility at pH values above its pI. Currently, experiments are on-going to do that.

3. Some theoretical explanations for protein-protein interactions in salt solutions

Recently Curtis and Lue wrote a comprehensive review of different theoretical treatments for understanding protein-protein interactions in salt solutions, pointing out that there is no single unified theoretical framework to rationalize the specificity of salt ion effects on protein intermolecular interactions (14). One of the important theories is the DLVO theory, in which proteins are treated as colloidal particles because their sizes are in the nanometer range (9). The DLVO theory was named after the scientists: Derjaguin and Landau, and Verwey and Overbeek (9). This theory lays the foundation for explaining the interparticle electrostatic interactions in low salt concentrations below 0.1 M in the most simplified way when the protein is net-charged. Specifically, the intermolecular interactions between two protein molecules in low salt concentrations can be described by the following equation (28):

$$w_2(r) = w_{ex}(r) + w_{disp}(r) + w_{elec}(r) \tag{1}$$

Where r is the center-to-center distance from two molecules; $w_{ex}(r)$ is the repulsive protein hard-sphere (excluded-volume) potential; $w_{disp}(r)$ is the attractive dispersion potential; $w_{elec}(r)$ is the electric double-layer repulsion potential, which can be further described by Debye-Huckel theory as the following:

$$w_{elec}(r) = \frac{(ze)^2 \exp[-\kappa(r-d)]}{4\pi r \varepsilon_0 \varepsilon_r (1 + \frac{\kappa d}{2})^2} \quad \text{for } r > \sigma \tag{2}$$

Where ze is the net charge of a protein, e is the elementary charge, ε_0 is the dielectric permittivity of vacuum, ε_r is the dielectric constant of water, and κ is the inverse Debye length calculated by

$$\kappa^2 = \frac{2e^2 N_A I}{kT \varepsilon_0 \varepsilon_r} \tag{3}$$

Where I is the ionic strength of the solution, k is the Boltzmann's constant, T is the absolute temperature, and N_A is the Avogadro's number.

As presented in Equation 2, it is obvious that the more net charges a protein carries, the stronger the electrostatic double-layer repulsive force becomes. Also, Equation 2 indicates the addition of the salts monotonically decreases (or screens) the double-layer repulsion, and then reaches a plateau (the exponential term approach zero). The general screening effect is consistent with the initial drop in protein solubility and rise in liquid-liquid phase transition temperature as described above for the charged proteins. The DLVO theory was used to explain the protein solubility decrease of lysozyme (23). It should be pointed out that it is difficult to differentiate between the direct binding of salt ions to their opposite-charged partners on the protein surface and the screening by the salt-ion layer near the protein surface. The reason is that the first type of interaction decreases the double layer repulsion through balancing out the "ze" term in Equation 2 while the second type of interaction work through κ, the inverse Debye length. One of the major limitations of the DLVO theory is lack of ion-specificity as presented in Equation 2 and both cation and anion contribute equally as far as they have the same valency. Therefore, the DLVO theory cannot explain the anion-specific modulations on protein-protein interactions, i.e. the direct or reverse Hofmeister series at pH 5.3 for the antibody (3). In addition, the DLVO theory suggests that the double-layer repulsion decreases and levels off with salt addition, in contrary to the numerous nonmonotonic behavior mentioned above in Historical Background.

For a charge-neutral species (i.e. proteins at their pI), many other theoretical considerations were developed to explain the initial salting-in and later salting-out behavior (19, 29, 30). They can be used to explain the general pattern of protein-protein interactions. In essence, the electrostatic interactions and hydrophobic interactions are the two major types of intermolecular forces (20, 31). The effects from the electrostatic interactions on the free energy of a protein in a low salt concentration solution may be described by Debye-Huckel theory in combination with Kirkwood's expression of the protein dipole moment as follows (20, 31):

$$\Delta G_{e.s.} = A - \frac{B(I^{1/2})}{1 + C(I^{1/2})} - DdI \qquad (4)$$

Where A, B, C and D are constants, I is the ionic strength of the solution, d is the dipole moment for the protein. This theory predicts the salting-in effect: as the ionic strength increases, protein solubility rises. This idea is consistent with the observations of salting-in of proteins near pI. The main limitation of this theory is that it does not consider ion-specificity.

The free energy change for a protein involving the hydrophobic interactions may be illustrated by the cavity theory as follows(20):

$$\Delta G_{cav} = \left[N * Area + 4.8N^{1/3}(\kappa^e - 1)V^{2/3}\right] \left(\frac{\partial \sigma}{\partial m_3}\right) m_3 \qquad (5)$$

where N is Avogadro's number, $Area$ is the surface area of a protein molecule, κ^e corrects the macroscopic surface tension of the solvent to molecular dimensions, V is the protein's molar volume, $\left(\frac{\partial \sigma}{\partial m_3}\right)$ is the molal surface tension increment of the salt, and m_3 is the molality of the salt. This cavity theory describes how much free energy is needed to form a cavity in the solution to accommodate a hydrophobic protein molecule. Therefore, the surface tension of the solution is an important parameter and its modulation by salts impacts protein solubility and therefore protein-protein interactions. It predicts that the addition of kosmotropic salts, which increase the solution surface tension, will result in the salting-out effect and effectively strengthening of attractive protein-protein interactions. Therefore, these salting-in and salting-out effects in combination modulate protein solubility and protein-protein interactions in salt solutions (20, 31). Specifically, near the pI the salting-in effect dominates initially (protein solubility increases) and the addition of salts disrupts attractive protein-protein interactions. Then, further increase in (kosmotropic) salt concentration results in strengthening attractive protein-protein interactions as the salting-out effect begins to dominate (protein solubility decreases).

4. Molecular mechanism for protein-ion interactions

The simple ions shown in Figure 1 have different sizes, diverse hydration properties and polarizabilities (32). The interaction strength between an ion and water molecule in comparison to that between water-water determine the ion hydration property: an ion is strongly hydrated when it interacts with water molecules more strongly than the water-water interaction while the opposite makes an ion less hydrated (33-36). Shown in Figure 4 is the ranking of hydration property for the selected salt ions. Specifically, the large and more polarizable anion, i.e. SCN-, is less hydrated while the small and less polarizable anion, i.e. F-, is strongly hydrated.

The law of matching water affinities is the hallmark theory for defining the interaction strength between salt ions and proteins thermodynamically, in which the hydration and size properties of the ions and their counterparts on the protein surface are the key for explaining the protein-protein interaction behavior (33-36). Specifically according to the law of matching water affinities, oppositely charged ions in solutions form inner sphere ion pairs spontaneously when they have similar water affinities (36).

The chemistry of protein surface is heterogeneous, composed of both positive and negative-charged residues, and polar and nonpolar groups. As shown in Figure 4, monovalent anions of SCN- and halides, except F-, were weakly hydrated because of their large size, in comparison to the small-size monovalent cations being reasonably hydrated. On the protein surface, the positive-charged side chains on Arg, Lys and His are all derivatives of ammonium and therefore they are all weakly hydrated, matching well with the weakly hydrated SCN-. According to the law of matching water affinity, the weakly hydrated anions, such as SCN-, have the strongest interactions with the positive-charged side chains from the protein and neutralize them, followed by Cl- and F-. On the other hand, the negative-charged side chains from Asp and Glu are strongly hydrated carboxylate, mismatching with Na^+ and K^+ whose interaction strengths are similar to that between water molecules (33-36). To the contrary, the divalent cation, i.e. Mg^{2+}, interacts with water molecules more strongly than Na^+ and K^+ and is strongly hydrated. It is then expected that the divalent cation interacts with the carboxylate more strongly than both Na^+ and K^+.

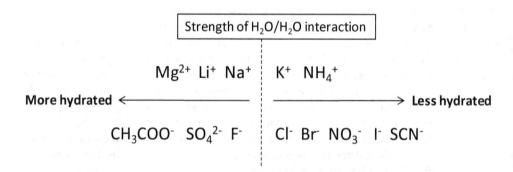

Fig. 4. Hydration properties of selected salt ions (34, 36).

Protein surface is composed of not only polar functional groups from the amide bonds of the exposed peptide backbone and the side chains of Asn and Gln, but also non-polar functional groups from the side chains of Phe, Ile and other amino acids. Both the polar and non-polar groups can be considered as weakly hydrated (37). Collins proposed that the weakly hydrated anions could also interact with both of the groups, besides the charged side chains (33-36). Recently, it was demonstrated, through a molecular dynamics (MD) study of lysozyme in a mixed aqueous solution of potassium chloride and iodide (0.4 M), that weakly hydrated anions, i.e. I-, preferred to interact with the nonpolar groups besides the positive-charged residues on lysozyme (38). Furthermore, the interaction between weakly hydrated anions and the amide bonds was also proposed based on the solubility study on poly(N-isopropylacrylamide) in salt solutions (39). For cations, it has been shown that both Ca^{2+} and Mg^{2+} can interact strongly with proteins through the diopolar amide bond (40) (18, 41).

The electroselectivity theory deserves attention when considering salt ion-protein interactions. Developed based on the anions' affinity for the anion exchanger, the electroselectivity theory proposed, purely based on the electrostatic interaction, that the ions with higher valency, such as SO_4^{2-}, interact with the positive-charge residues on the protein surface more strongly than those with a single valence, such as SCN^-(42, 43). The strong electrostatic interactions imparted by SO_4^{2-} were recently demonstrated by exploring specific ion effects on interfacial water structure adjacent to a bovine serum albumin at pH conditions below its pI using vibrational sum frequency spectroscopy (VSFS) (44).

5. From protein-ion interactions to protein-protein interactions

The complexity of protein-protein interactions as modulated by salt ions at low concentrations might be explained from the framework of dominance of specific electrostatic interactions from both cation and anions for the protein surface, concomitantly considering the following biophysical properties including net charge, surface charge density and hydration of a protein, and hydration, size, polarizability and valency of salt ions.

The first key property is the macroscopic net charge (considering the protein as a particle) as modulated by pH. First, a protein is net charge neutral, positively-charged and negatively-charged at pH near, below, or above its pI, respectively. Furthermore, patches of protein surface could be macroscopically weakly-hydrated because of the abundantly exposed nonpolar and polar groups, regardless of whether a protein surface is overall hydrophobic or hydrophilic. It was pointed out that in general 1/3 of the protein surface is hydrophobic, resulting in a partially weakly-hydrated surface(45). Although the net charge of the protein is dictated by the solution pH, its nonpolar or polar surface might maintain its property of weak hydration when the native folding structure is not drastically affected by pH and low salt concentrations. As pH decreases below its pI, the increasingly net positive-charges, from the weakly hydrated side chains of Arg, His and Lys, might render the protein surface even more weakly hydrated. At pH above its pI, the strongly hydrated carboxylates, from the strongly hydrated side chains of Asp and Glu, bring more water onto the protein surface, which results in the surface becoming more hydrated.

5.1 pH near pI

A protein is net charge neutral at pI with the equal number of positive and negative-charged residues. Therefore the protein molecules may approach each other and fully explore complementary interaction configurations (46). It is well-known that a protein has the lowest solubility near its pI and easily precipitates, suggesting the presence of strong intermolecular attractive interactions. The interactions can be highly anisotropic due to ionic-pair interactions, cation-π interaction, hydrophobic interaction and others types of interactions. It is difficult to dissect which type of interaction contributes most to the intermolecular interactions, which might be sequence dependent and protein-specific.

Our previous experiment of antibody liquid-liquid phase separation near its pI suggests that the intermolecular interactions were attractive and sensitive to salts, indicating that there were electrostatic interactions between the antibodies. Our observations of the general salting-in trends in the solubility measurement and disruption of intermolecular electrostatic attractive interactions in the LLPS are in agreement of the solubility data at low salt concentrations for other proteins near their respective pI, i.e. carboxyhemoglobin (47). The idea of attractive electrostatic interactions is especially supported by the salting-in behavior near its pI by KF. Typically, KF only salts out neutral peptides without charged side chains and nonpolar small molecules (16, 17). The general salting-in trend is also consistent with the electrostatic interaction theory as described by Equation 4. However, this theory cannot explain the ranking of the anion's effectiveness for raising the antibody solubility.

In the monovalent K^+ salt solutions, K^+ does not match well with the strongly hydrated carboxylate as discussed above. In contrast, the water affinity of the weakly hydrated positive-charge side chains, polar and nonpolar groups match well with those weakly hydrated anions from SCN^- to Cl^-. It is then expected that K^+ interacts with protein surface fairly weakly and anion could specifically binds to the protein surface in which their specificities are determined by their binding constants for the protein. This idea is consistent with the specific anion's effect, as described by a direct Hofmeister series, of raising the antibody solubility and disruption of the intermolecular attractive interactions at pH 7.1. In addition, this idea is in agreement with the recent findings where a chaotropic monovalent anion bound more strongly to a net-charge neutral macromolecule, like BSA near its pI and polar Poly-(N-isopropylacrylamide), than a kosmotropic monovalent anion(44) (48).

On the other hand, strongly hydrated multivalent cation, such as Mg^{2+} and Ca^{2+}, could bind to the strongly-hydrated carboxylate. In addition, there are strong interactions between the amide bond and multivalent cation (17). The above two modes of binding could make multivalent cations strong salting-in reagents (just like the anions) at low salt concentrations, overshadowing the possible salt-outing of the nonpolar residues on a protein by the multivalent cations.

In short, the electrostatic attractive interactions may dominate at protein-protein interactions in low salt solutions at pH near its pI, where the binding strengths between the protein surface for both cation and anions, working in synergy, determines the salting-in effectiveness of the salts as they are initially added.

5.2 pH below pI

When a protein is net charged at pH above and below its pI, the aforementioned observations of protein-protein interactions initially becoming more attractive or drop in protein solubility suggest that (i) the electrostatic repulsion dominates the protein-protein interactions and (ii) the initial addition of the salts to a charged protein effectively neutralizes the net charge of the protein and reduces the electrostatic repulsion.

Below pI, the positive-charges on proteins are from the weakly hydrated side chains of Arg, Lys or His. In addition, polar and nonpolar sites on the protein surface are also weakly hydrated. As results, the more weakly hydrated a monovalent anion is, the more strongly it interacts with the positive-charged protein, and the more effectively it neutralizes the protein's net charge. The monovalent anions then follow the reverse Hofmeister series for their effectiveness of weakening the electrostatic repulsive intermolecular interactions and decreasing the protein solubility. This idea is consistent with the solubility measurement and phase transition data for both lysozyme and the antibody. The ranking for the binding strength between the anions and this antibody is also in agreement with what has been observed in monovalent salt solutions for other positive-charged proteins including other antibodies, BSA and lysozyme(49) (44) (22, 50). The binding of SO_4^{2-} to the positive-charged lysozyme and BSA, consistent with the electroselectivity theory, provides convincing experimental evidence that there is strong electrostatic interaction between a positive-charged protein and divalent anions, despite the mismatching water affinity.

The competitive interactions of co-ions against the counter-ions for a positive-charged protein become apparent for the strongly hydrated multivalent cation, i.e. Mg^{2+}. For example, Mg^{2+} may interact strongly at the strongly hydrated carboxylate or peptide groups in comparisons to Na^+ and K^+, effectively raising the positive-charges of the protein and hindering the anion's charge neutralization effect. Then, it appears that $MgCl_2$ will be less effective at weakening the electrostatic repulsive interactions and decreasing the protein solubility than NaCl (with the same molar concentration of Cl^-). Therefore, the protein-protein interactions are expected to be more repulsive in the $MgCl_2$ solutions than in the NaCl solutions, following the direct Hofmeister series. This notion is in agreement with the measurement of the phase transition temperature for lysozyme(21). Similarly, solubility of lysozyme in multivalent cation salt solutions was higher than that in the monovalent cation salt solutions with the same anion(24).

When anions complete their charge neutralization process as suggested by the minimum of protein solubility in Figure 3, the protein can be considered as pseudo charge-neutral. The salt's effect on protein-protein interactions then is expected to follow the direct Hofmeister series, as described above for a protein near its pI. This is the reason for why we observed the nonmonotonic behavior in the aforementioned proteins at pH below their pI.

5.3 pH above pI

On the other hand, at pH above its pI, the protein is negatively charged. Although the net negative charges are from the strongly hydrated carboxylate side chains on Asp and Glu, its surface still has significant presence of polar and nonpolar residues, attracting weakly hydrated anions. It is anticipated that the competitive bindings of cation and anion for protein surface determine the final effect on protein-protein interactions and solubility. The counterions with strong electrostatic interactions with the proteins, i.e. multivalent cations, can neutralize the net charge, weaken the repulsive electrostatic intermolecular interactions and decrease the protein solubility more effectively than the monovalent cations of Na^+, following the reverse Hofmeister series. Furthermore, in the Na^+ salt solutions, the anion's binding to the weakly hydrated sites, possibly stronger than that between Na^+ and the carboxylate, may effectively increase the repulsive interactions. This is consistent with the experimental observation of the experimental findings for protein-protein interactions of ovalbumin in NaCl and YCl_3 solutions at pH conditions above its pI. Specifically, in the NaCl solution Cl's binding to ovalbumin preempted that of Na^+, effectively raising the intermolecular repulsive interactions. On the other hand, the trivalent Y^{3+} could bind to the carboxylate strongly, neutralize the net negative-charges and weaken the repulsive intermolecular interactions. After charge neutralization, the salting-in effect by YCl_3 followed.

However, when either strongly hydrated F- or acetate was used, they mismatched for both the positive-charged side chains and weakly hydrated polar and nonpolar residues on the net negative-charged protein surface. Possibly, Na^+ now might interact with the protein stronger than F- or acetate and neutralize the negative charges. This could be a reasonable explanation for the nonmonotonic behavior mentioned for Apoferrtin in NaAcetate solution, but not in the NaCl solution.

5.4 Surface charge density

The surface charge density of a protein could dramatically change the above nonmonotonic behavior. At pH close to the pI or a large-size protein with small number of either positive or negative net charges, where the surface charge density is low, only the monotonic salting-in behavior could be observed because the charge neutralization process is less dramatic. On the other hand, when a protein has high surface charge density due to either a small size or a large number of positive charges, the anions might not completely neutralize the positive charges even at molar concentration and therefore only a decrease in protein solubility can occur. As a matter of fact, this might be for the case of lysozyme solubility at pH 4 and 7, especially when a weak chaotropic anion, i.e. Cl-, was used(22). The reason is that Cl- could bind to the protein surface less strongly and effectively at weakening the electrostatic repulsive interactions than a strong chaotropic anion, such as SCN-. But at pH 9.4 where the surface charge density was smaller than at pH 4 and 7, the weakly hydrated SCN- could neutralize the net charges completely, and as a result the nonmonotonic behavior appeared.

As proteins transition from a high surface charge density system to low, the interaction between a co-ion and charged surface could be explained through the smeared surface charge model and discrete surface charge model, respectively. In a low surface charge density system (discrete charge surface), such as a large-size antibody, the co-ion binding probably becomes more significant, in comparison to a small globular protein, i.e. lysozyme, of a high surface-charged density system. The reason is that the co-ion can approach the surface without experiencing the repulsive electrostatic force. This idea of co-ion adsorption to a low or medium negative-charged hydrophobic surface is supported by the recent molecular simulation for a self-assembled monolayer (51). The simulation results shows that even at a high surface charge density of – 2.0×10^{-2} C/m^2, there was significant co-ion adsorption. Therefore, significant presence of co-ion adsorption is expected for a typical protein surface with a surface charge density in the low range of mC/m^2 (10, 52),.

5.5 Additional attractive interaction by polarizable anions

Another important feature of protein-protein interactions in salt solutions is the presence of possible additional protein-protein attractive force caused by the weakly hydrated anions for a positive-charged protein, although the exact mechanism remains to be defined. A recent Monte Carlo simulation reveals that the presence of chaotropic (or polarizable) ions, like SCN⁻, introduced this additional interaction of dispersion force in nature between protein molecules (53). More importantly, liquid-liquid phase separation of the antibody at different pHs in a KSCN solution at a pH below its pI indicates that this attractive protein-protein interaction became stronger as the pH dropped and the protein carried more positive charges.

6. Conclusions

Despite the complexity of salt ion and protein interactions and their effects on protein-protein interactions, the rich salt-specific effect at low salt concentrations may be qualitatively explained based on the specific binding of both anions and cations for protein surface with heterogeneous surface chemistry as illustrated in Figure 5. In the future, it would be beneficial to have a quantitative description for the salt ions' effect on protein-protein interactions.

As shown in Figure 5, protein surface may always have hydrophobic patches, which are weakly hydrated and matches well with the weakly hydrated anions. Additionally, the exposed dipolar amide bond of the peptide backbone is the potential site for the divalent cation and weakly hydrated anions. Furthermore, pH change not only modulates the net charge property of the protein but also modifies the degree of surface hydration. Specifically, as the pH decreases away from their pIs, proteins become net positively-charged and even more weakly hydrated because the positive-charges are from the weakly hydrated side chains of Arg, Lys, and His. At pH values close to their pIs, proteins are net-charge neutral. Then as pH increases away from their pI, proteins become becomes net negatively-charged and less weakly hydrated because the negative charges are from strongly hydrated carboxylate from Asp and Glu.

At a pH close to the pI of a protein, both cations and anions can access the neutral protein and may work in synergy to disrupt the attractive intermolecular protein interactions and result an increase of protein solubility. On the other hand, they work competitive for a sufficiently charged protein (in Figure 5). Specifically, the counter-ion from the salt tends to neutralize the net charge of the protein, weakening the electrostatic repulsive intermolecular interactions while the co-ion is likely to hinder the charge-neutralization effect by the counter-ion, effectively strengthening the repulsive intermolecular interactions. The interaction strength between the ions and protein surface is dependent on both electrostatic and hydration properties for both ions and protein. The final outcome of protein-protein interactions is then determined by a combination of the protein surface charge density and the relative binding strength of both ions for the protein surface. When the counter-ions interact with the charge protein more strongly than the co-ions, the charge neutralization step dominates, resulting in protein-protein interactions becoming less repulsive, after which there could be the salting-in effect as if the protein-counter-ion complex is pseudo charge-neutral. In the opposite situation, the strong interaction from the co-ions effectively renders the protein-protein interactions more repulsive.

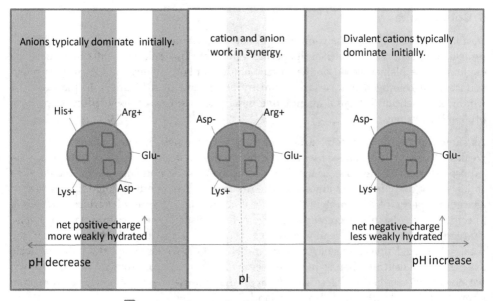

Fig. 5. Schematic illustration of the changes in net charge and hydration properties of a protein as pH varies.

7. Acknowledgements

The author would like to thank Dr. Izydor Apostol for reviewing the manuscript and providing valuable suggestions.

8. References

[1] Wei, W. 2005. Protein aggregation and its inhibition in biopharmaceutics. International Journal of Pharmaceutics 289:1-30.

[2] Schmidt, S., D. Havekost, K. Kaiser, J. Kauling, and H. J. Henzler. 2005. Crystallization for the Downstream Processing of Proteins. Engineering in Life Sciences 5:273-276.

[3] Chayen, N. E., editor. 2007. Protein Crystallization Strategies for Structural Genomics International University Line

[4] Gunton, J. D., A. Shiryayev, and D. L. Pagan. 2007. Protein condensation: kinetic pathways to crystallization and disease. Cambrige University Press, Cambridge.

[5] Uversky, V., and A. L. Fink. 2006. Protein Misfolding, Aggregation, and Conformational Disease Part A: Protein Aggregation and Conformational Diseases. Springer Science+ Business Media, Inc., Singapore.

[6] Kunz, W., editor. 2010. Specific Ion Effects. World Sceintific Publishing Co., Singapore.

[7] Zangi, R. 2010. Can salting-in/salting-out ions be classified as chaotropes/kosmotropes? J. Phys. Chem. B 114:643-650.

[8] Erickson, H. P. 2009. Size and Shape of Protein Molecules at the Nanometer Level Determined by Sedimentation, Gel Filtration, and Electron Microscopy. Biol Proced Online 11:32-51.

[9] Israelachvili, J. 1991. Intermolecular & Surface Forces. Academic Press, London.

[10] Mason, B. D., J. Zhang-van Enk, L. Zhang, R. L. Remmele, and J. Zhang. 2010. Liquid-liquid phase separation of a monoclonal antibody and influence of Hofmiester anions. Biophysical Journal 99:3792-3800.

[11] Zhang, L., Tan, H., Fesinmeyer, R. Matthew, Li, C., Catrone, D., Le, D., Remmele, R.L., and Zhang, J. 2011. Antibody Solubility Behavior in Monovalent Salt Solutions Reveals Specific Anion Effects at Low Ionic Strength J. Pharm. Sci. accepted.

[12] Broide, M. L., T. M. Tomine, and M. D. Saxowsky. 1996. Using phase transitions to investigate the effect of salts on protein interactions. Physical Review E 53:6325-6335.

[13] Finet, S., D. Vivarès, F. Bonneté, A. Tardieu, Charles W. Carter, Jr., and M. S. Robert. 2003. Controlling Biomolecular Crystallization by Understanding the Distinct Effects of PEGs and Salts on Solubility. In Methods in Enzymology. Academic Press. 105-129.

[14] Curtis, R. A., and L. Lue. 2006. A molecular approach to bioseparations: protein-protein and protein-salt intractions. Chemical Engineering Science 61:907-923.

[15] Ries-Kautt, M. M., and A. F. Ducruix. 1989. Relative effectiveness of various ions on the solubility and crystal growth of lysozyme. J. Biol. Chem. 264:745-748.

[16] Robinson, D. R., and W. P. Jencks. 1965. The effect of concentrated salt solutions on the activity coefficient of acetyltetraglycine ethyl ester. J. Am. Chem. Soc. 87:2470-2479.

[17] Nandi, P. K., and D. R. Robinson. 1972. Effects of salts on the free energies of nonpolar groups in model peptides. Journal of the American Chemical Society 94:1308-1315.

[18] Nandi, P. K., and D. R. Robinson. 1972. The effects of salts on the free energy of the peptide group. J. Am. Chem. Soc. 94:1299-1308.

[19] Baldwin, R. L. 1996. How Hofmeister ion interactions affect protein stability. Biophysical Journal 71:2056-2063.

[20] Arakawa, T., Timasheff, S. N. 1985. Theory of protein solubility. Meth. Enzymol. 114:49-77.

[21] Grigsby, J. J., H. W. Blanch, and J. M. Prausnitz. 2001. Cloud-point temperatures for lysozyme in electrolyte solutions: effect of salt type, salt concentration and pH. Biophysical Chemistry 9:231-243.

[22] Zhang, Y., and P. S. Cremer. 2009. The inverse and direct Hofmiester series for lysozyme. Proc. Natl. Acad. Sci USA 106:15249-15253.

[23] Retailleau, P., M. Ries-Kautt, and A. Ducruix. 1997. No salting-in of lysozyme chloride observed at low ionic strength over a large range of pH. Biophysical Journal 73:2156-2163.

[24] Benas, P., L. Legrand, and M. Riess-Kautt. 2002. Strong and specific effects of cations on lysozyme chloride solubility. Acta Cryst D 58:1582-1587.

[25] Ianeselli, L., F. Zhang, M. W. A. Skoda, R. M. J. Jacobs, R. A. Martin, S. Callow, S. Prei vost, and F. Schreiber. 2010. Proteinâˆ'Protein Interactions in Ovalbumin Solutions Studied by Small-Angle Scattering: Effect of Ionic Strength and the Chemical Nature of Cations. The Journal of Physical Chemistry B 114:3776-3783.

[26] Zhang, F., M. W. A. Skoda, R. M. J. Jacobs, S. Zorn, R. A. Martin, C. M. Martin, G. F. Clark, S. Weggler, A. Hildebrandt, O. Kohlbacher, and F. Schreiber. 2008. Reentrant Condensation of Proteins in Solution Induced by Multivalent Counterions. Physical Review Letters 101:148101.

[27] Petsev, D. N., B. R. Thomas, S. T. Yau, and P. G. Vekilov. 2000. Interactions and Aggregation of Apoferritin Molecules in Solution: Effects of Added Electrolytes. Biophysical Journal 78:2060-2069.

[28] Liu, W., D. Bratko, J. M. Prausnitz, and H. W. Blanch. 2004. Effect of Alcohols on Aqueous Lysozyme-Lysozyme Interactions from Static Light-Scattering Measurements. Biophysical Chemistry 107:289-298.

[29] Timasheff, S. N. 1985. Theory of protein solubility. Meth. Enzymol. 114:49-77.

[30] Melander, W., and C. Horvath. 1977. Salt effect on hydrophobic interactions in precipitation and chromatography of proteins: an interpretation of the lyotropic serie. Arch Biochem Biophys 183:200-215.

[31] Tanford, C., 1966. Physical Chemistry of Macromolecules. John Wiley & Sons, Inc., New York.

[32] Collins, K. D., and M. W. Washabaugh. 1985. The Hofmeister effect and the behaviour of water at interfaces. Quarterly Reviews of Biophysics 18:323-422.

[33] Collins, K. D., G. W. Neilson, and J. E. Enderby. 2007. Ions in water: characterizing the forces that control chemical processes and biological structure. Biophysical Chemistry 128:95-104.

[34] Collins, K. D. 2006. Ion hydration: Implications for cellular function, polyelectrolytes, and protein crystallization. Biophysical Chemistry 119:271-281.

[35] Collins, K. D. 2004. Ions from the Hofmiester series and osmolytes: effect on proteins in solution and in the crystallization process. Methods 34:300-311.

[36] Collins, K. D. 1997. Charge density-dependent strength of hydration and biological structure. Biophysical Journal 72:65-75.

[37] Collins, K. D. 1995. Sticky ions in biological systems. Proceedings of the National Academy of Sciences 92:5553-5557.

[38] Lund, M., L. Vrbka, and P. Jungwirth. 2008. Specific Ion Binding to Nonpolar Surface Patches of Proteins. Journal of the American Chemical Society 130:11582-11583.

[39] Zhang, Y., S. Furyk, D. E. Bergbreiter, and P. S. Cremer. 2005. Specific Ion Effects on the Water Solubility of Macromolecules: PNIPAM and the Hofmeister Series. Journal of the American Chemical Society 127:14505-14510.

[40] Baldwin, R. L. 1996. How Hofmeister ion interactions affect protein stability. Biophysical Journal 71:2056-2063.

[41] Zhang, Y., and P. S. Cremer. 2006. Interactions between macromolecules and ions: the Hofmiester series. Curr. Opin. Chem. Biol. 10:658-663.

[42] Gjerde, D. T., G. Schmuckler, and J. S. Fritz. 1980. Anion Chromatography with low-conductivity eluents. II. Journal of Chromatography A 187:35-45.

[43] Gregor, H. P., J. Belle, and R. A. Marcus. 1954. Studies on Ion Exchange Resins. IX. Capacity and Specific Volumes of Quaternary Base Anion Exchange Resins1. Journal of the American Chemical Society 76:1984-1987.

[44] Chen, X., S. C. Flores, S.-M. Lim, Y. Zhang, T. Yang, J. Kherb, and P. S. Cremer. 2010. Specific Anion Effects on Water Structure Adjacent to Protein Monolayers Langmuir 26:16447-16454.

[45] Schwierz, N., D. Horinek, and R. R. Netz. 2010. Reversed Anionic Hofmeister Series: The Interplay of Surface Charge and Surface Polarity. Langmuir 26:7370-7379.

[46] Leckband, D., and J. Israelachvili. 2001. Q. Rev. Biophys.

[47] Green, A. 1932. Studies in the physical chemistry of the proteins X: the solubility of hemoglobin in solutions of chlorides and sulfates of varying concentration. J. Biol. Chem. 95:47-66.

[48] Chen, X., T. Yang, S. Kataoka, and P. S. Cremer. 2007. Specific Ion Effects on Interfacial Water Structure near Macromolecules. Journal of the American Chemical Society 129:12272-12279.

[49] Fesinmeyer, R., S. Hogan, A. Saluja, S. Brych, E. Kras, L. Narhi, D. Brems, and Y. Gokarn. 2009. Effect of Ions on Agitation- and Temperature-Induced Aggregation Reactions of Antibodies. Pharmaceutical Research 26:903-913.

[50] Gokarn, Y. R., R. M. Fesinmeyer, A. Saluja, V. Razinkov, S. F. Chase, T. M. Laue, and D. N. Brems. 2011. Effective charge measurements reveal selective and preferential accumulation of anions, but not cations, at the protein surface in dilute salt solutions. Protein Science 20:580-587.

[51] Lima, E. R. A., M. Bostrom, D. Horinek, E. C. Biscaia, W. Kunz, and F. W. Tavares. 2008. Co-Ion and Ion Competition Effects: Ion Distributions Close to a Hydrophobic Solid Surface in Mixed Electrolyte Solutions. Langmuir 24:3944-3948.

[52] Sivasankar, S., S. Subramaniam, and D. Leckband. 1998. Direct molecular level measurements of the electrostatic properties of a protein surface. Proceedings of the National Academy of Sciences 95:12961-12966.

[53] Tavares, F. W., D. Bratko, H. W. Blanch, and J. M. Prausnitz. 2004. Ion-specific effects in the colloid-colloid or protein-protein potential of mean force: role of salt-macroion van der Waals interactions. J. Phys. Chem. B. 108:9228-9235.

3

NMR Investigations on Ruggedness of Native State Energy Landscape in Folded Proteins

Poluri Maruthi Krishna Mohan
Department of Chemistry & Chemical Biology,
Rutgers University, New Jersey,
USA

1. Introduction

The ability of proteins to adopt their functional, highly structured states in the intracellular environment during and after its synthesis is one of the most remarkable evolutionary achievements of biology. Deciphering the code of protein self-organization process has been an intellectual challenge for scientists over the past few decades. Although the structure-function paradigm about folded structures and functions remains valid, the role of internal dynamics and conformational fluctuations in protein function is becoming increasingly evident (Bhabha *et al* 2011; Boehr *et al* 2006; Eisenmesser *et al* 2005; Fraser *et al* 2009; Mittermaier and Kay 2006; Parak 2003b; Popovych *et al* 2006; Tzeng and Kalodimos 2009; Whitten *et al* 2005). Further, recent structural and genomic data have clearly shown that not all proteins have unique folded structures under normal physiological conditions. Hence, the way a protein exists, is bound to have a profound effect on its function.

A complete understanding of protein folding process requires characterization of all the species populating along the folding coordinate, these include the unfolded state, the partially folded intermediate states, low energy excited states and the fully folded native state. The most recent and widely accepted model is the 'funnel view' of protein folding (Bryngelson *et al* 1995; Dill and Chan 1997; Onuchic *et al* 1997; Shoemaker *et al* 1999; Wolynes 2005) also known as the 'Energy landscape model' **(Fig. 1)**, which is inclusive of the earlier concepts of 'folding pathways'. According to this model protein folding is a parallel, diffusion-like motion of conformational ensemble on the energy landscape biased towards the native state. This model is free from the Levinthal paradox (Dill and Chan 1997; Levinthal C 1969) as it envisages the process of reaching a global minimum in free energy as a rapid process occurring by multiple routes on a funnel like energy landscape **(Fig. 1)**. This view focuses on the rapid decrease of the conformational heterogeneity in the course of the folding reaction and is based on a statistical description of a protein's potential surface (Wolynes *et al* 1995; Wolynes 2005). The depth in the funnel represents the free energy of the polypeptide chain in fixed conformations and the width indicates the chain entropy **(Fig. 1)**. The funnel becomes narrower in the lower energy region because of the low chain entropy. The broad end of the funnel reflects the heterogeneous unfolded state, while the narrow end represents the supposedly homogeneous native state (Dill and Chan 1997; Dobson and Karplus 1999; Dyson and Wright 2005). Different members of the ensemble may fold/unfold along independent pathways and their energy profiles could be different. Protein folding theories start from the unfolded state **(Fig. 1)** and encompass a range of topologies like the pre-molten globule, the molten globule and various other ordered or disordered forms as the protein folds down the funnel (Dunker *et al* 2002; Uversky 2002; Uversky 2003).

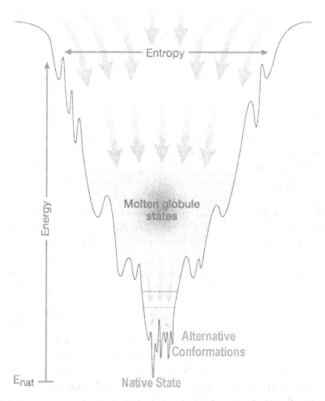

Fig. 1. A schematic energy landscape view of protein folding: The surface of the funnel represents a whole range from the multitude of denatured conformations to the unique native structure (Dill and Chan 1997). The ordered state is the natively folded structure of a protein that has a well defined secondary and tertiary structure. Alternative conformations are higher energy native state conformations and contain all the secondary and tertiary structural characteristics of folded state. Molten globule states are intermediates in the protein folding pathway with compact structures that exhibit a high content of secondary structure, nonspecific tertiary structure, and significant structural flexibility. Random coils are highly unstructured protein denatured states.

In a living cell, a polypeptide chain chooses between three potential fates - functional folding, potentially deadly misfolding and mysterious non-folding (Dobson 2003). This choice is dictated by the peculiarities of amino acid sequence and/or by the pressure of environmental factors. The biological function of a protein arises as a result of interplay between specific conformational forms, namely, native state (ordered forms), low energy excited states, molten globules, pre-molten globules, and denatured state (random coils). In view of this, it will not be an exaggeration to assume an ensemble existence of all these states at any particular time, their relative abundance being governed by basic thermodynamics. Upon ligand binding or some signaling modification, concentration of one state may increase at the expense of the others. This can explain the fast regulatory steps involved in various biological functions.

Much of structural biology of proteins is so focused on studies of native state, providing detailed atomic descriptions and coordinates of static three-dimensional (3D) structures. A large body of evidence using a diverse spectrum of biophysical methods clearly establishes that proteins are dynamic over a broad range of timescales and such dynamics play critical roles in various biological processes, such as: initial formation of encounter complexes in macromolecular association, target searching in specific protein-protein/protein-DNA recognition, conformational preferences in ligand binding, conformational transitions associated with allostery, the course of enzyme catalysis, intermediates along the protein folding pathway, and early events in self-assembly processes (Bai *et al* 1995; Boehr *et al* 2006; Clore 2011; Dunker *et al* 2002; Eisenmesser *et al* 2005; Feher and Cavanagh 1999; Fraser *et al* 2009; Kitahara *et al* 2005; Korzhnev *et al* 2003; Kumar *et al* 2007; Lambers *et al* 2006; Mohan *et al* 2006; Piana *et al* 2002; Popovych *et al* 2006; Tang *et al* 2008; Villali and Kern 2010). The amplitudes and the timescales of motion that characterize the dynamics of a protein under a given set of conditions can be understood in terms of an 'energy landscape' as described above. Ground-state conformers that occupy the bottom of the energy landscape funnel and are separated from other conformational states by very small kinetic barriers that are easily overcome by thermal energy form the basis of structural studies by NMR (Nuclear Magnetic Resonance Spectroscopy) and X-ray diffraction for last few decades.

In general, the dynamic phenomenon involves the inter-conversion between ground state conformers with higher energy structures known as 'excited states'. The populations of these low-energy excited states/near native states/alternative conformations at equilibrium are very sparse and their lifetimes are short. Moreover, these transient states arising from rare but rapid excursions between the global free energy minimum and higher free energy local minima are extremely challenging to study at atomic resolution under equilibrium conditions since they are effectively invisible to most structural and biophysical techniques including crystallography and conventional NMR spectroscopy (Bhabha *et al* 2011; Boehr *et al* 2006; Eisenmesser *et al* 2005; Fraser *et al* 2009; Mittermaier and Kay 2006; Popovych *et al* 2006; Tzeng and Kalodimos 2009; Whitten *et al* 2005; Clore and Iwahara 2009; Clore 2011). However, a complete understanding of the conformational fluctuations these bio-molecules undergo is essential to gain an insight into their biochemical and biophysical properties. Hence, it is critical to characterize the structural ensembles that describe these functionally important states and the mechanisms by which they interconvert with the ground-state conformers.

Recent developments in NMR, however, have rendered short-lived, sparsely populated states accessible to spectroscopic analysis, yielding considerable insights into their kinetics, thermodynamics, and structures. Over the past decade, new and powerful NMR approaches such as paramagnetic relaxation enhancement (PRE) (Clore and Iwahara 2009; Clore 2011), relaxation dispersion (RD) (Boehr *et al* 2006; Mittermaier and Kay 2006; Tzeng and Kalodimos 2009) and non-linear temperature dependence of amide proton chemical shifts (Krishna Mohan *et al* 2008; Mohan *et al* 2008b; Tunnicliffe *et al* 2005; Williamson 2003) have emerged and significantly contributed to our understanding of the relationship between structure, dynamics and function of proteins with respect to the excited-state conformers that are sparsely populated and often exist transiently.

In the present chapter the theoretical basis of NMR approach for the curved temperature dependence of amide proton chemical shifts will be discussed in detail. Theoretical simulations will be presented to understand the nature and extent of curvature of the chemical shifts. Further, experimental studies performed till date on different protein systems will be reviewed to demonstrate the curved temperature dependence of amide proton chemical shifts as a tool to detect the low populated near native states/ alternative conformations of the protein residues. Moreover, the significance of these conformational fluctuations will be evaluated with regard to protein function and folding.

2. Theory of curved temperature dependence of amide proton chemical shifts

NMR chemical shift is a sensitive indicator of the environment and molecular conformation. In proteins ^{1}H, ^{13}C and ^{15}N chemical shifts are sensitive to protein secondary structures and are used to deduce the preliminary structural information (Schwarzinger *et al* 2000; Schwarzinger *et al* 2001; Wishart and Sykes 1994; Wishart *et al* 1995; Wüthrich K 1986). The temperature dependence of amide proton chemical shifts in globular proteins has been investigated for over more than three decades by many researchers and continues to be investigated even today (Anderson *et al* 1997; Baxter and Williamson 1997; Cierpicki and Otlewski 2001; Krishna Mohan *et al* 2008). The amide proton chemical shifts are directly proportional to bond magnetic anisotropy (σ^{ani}) and this is crucially dependent on H-bonding, either intramolecular or intermolecular. In the former case, the carbonyl groups, the H-bond acceptors play a crucial role. The bond magnetic anisotropy is proportional to r^{-3} where r is the distance between the affected amide proton and the centre of the bond magnetic anisotropy, which lies close to the oxygen atom in the carbonyl groups (Krishna Mohan *et al* 2008). In case of solvent accessible groups, H-bonding with solvent molecules influences the amide proton chemical shifts. Thus the amide proton chemical shift is critically dependent on the length of the H-bond the proton is engaged in.

When the temperature of the solution is raised, thermal fluctuations increase which results in an increase in the average distance between atoms; X-ray crystallographic studies at several temperatures (98 – 320 K) on ribonuclease-A indicated that the protein volume increases linearly with temperature to an extent of about 0.4% per 100 K (Tilton, Jr. *et al* 1992). Such an increase in the distance between the atoms participating in a H-bond results in weakening of the H-bond. Consequently, chemical shifts of most amide protons move up field when the temperature is increased. Since bond magnetic anisotropy (σ^{ani}) is proportional to r^{-3} and molecular volume (V) is proportional to r^3, there is an inverse relationship of their variation with temperature ((σ^{ani}) α $1/V$). However, over a small temperature range, (σ^{ani}) may appear to decrease linearly with temperature, and

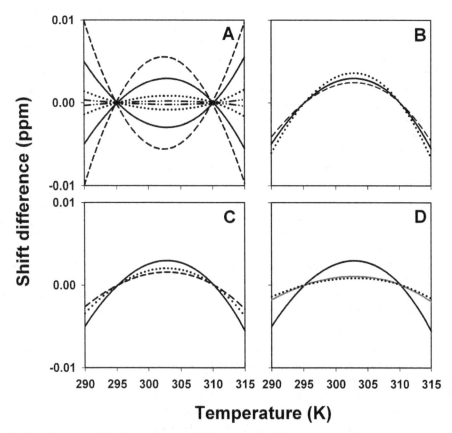

Fig. 2. Simulations of the dependence of H^N chemical shift variation with temperature (290 K – 315 K). In all the calculations shown chemical shifts are calculated and fitted to a straight line. Then the deviations from linearity are used to derive the residual curvatures (Krishna Mohan *et al* 2008; Williamson 2003). **(A)** Different curves show the dependence on free energy difference between the native and the higher energy alternate state: $\Delta G = 1$ kcal/mol (dashed), $\Delta G = 2$ kcal/mol (solid), $\Delta G = 3$ kcal/mol (dotted) and $\Delta G = 4$ kcal/mol (dash double dot dash); for these $T\Delta S$ at 298 K was fixed at 5.1 kcal/mol and ΔH varied as 6.1, 7.1, 8.1, and 9.1 kcal/mol respectively. The chemical shift and gradient parameters are: $\delta_1 = 8.5$ ppm, $\delta_2 = 8.0$ ppm, and $g_1 = -2$ ppb/K, $g_2 = -7$ ppb/K for convex shapes and $\delta_1 = 8.0$ ppm, $\delta_2 = 8.5$ ppm, and $g_1 = -7$ ppb/K, $g_2 = -2$ ppb/K for concave shapes. **(B)** The solid curve is the same as in '**A**' ($\Delta G = 2$ kcal/mol); Dashed curve, $\delta_1 = 8.1$ ppm, $\delta_2 = 8.0$ ppm, $g_1 = -2$ ppb/K, $g_2 = -7$ ppb/K, $\Delta G = 2$ kcal/mol; dotted curve, $\delta_1 = 9.0$ ppm, $\delta_2 = 8.0$ ppm, $g_1 = -2$ ppb/K, $g_2 = -7$ ppb/K, $\Delta G = 2$ kcal/mol **(C)** The solid curve is the same as in '**A**' ($\Delta G = 2$ kcal/mol); dashed curve, $\delta_1 = 8.5$ ppm, $\delta_2 = 8.0$ ppm, $g_1 = -2$ ppb/K, $g_2 = -4$ ppb/K, $\Delta G = 2$ kcal/mol; dotted curve, $\delta_1 = 8.5$ ppm, $\delta_2 = 8.0$ ppm, $g_1 = -4$ ppb/K, $g_2 = -7$ ppb/K, $\Delta G = 2$ kcal/mol. **(D)** The solid black curve ($\Delta G = 2$ kcal/mol) and the dotted curve ($\Delta G = 3$ kcal/mol) are the same as in '**A**'; solid grey curve is for $\delta_1 = 8.1$ ppm, $\delta_2 = 8.0$ ppm, $g_1 = -2$ ppb/K, $g_2 = -4$ ppb/K with $\Delta G = 2$ kcal/mol.

consequently amide proton chemical shifts would appear to vary linearly with temperature. In BPTI (basic pancreatic trypsin inhibitor) and lysozyme which are known to be extremely stable under a variety of extreme conditions, including temperature, it was indeed observed that the amide proton chemical shifts change linearly with temperature over the ranges, 279-359 K for BPTI and 278 – 328 K for Lysozyme (Baxter and Williamson 1997). Such measurements have been carried out on many other proteins (Cierpicki and Otlewski 2001; Cierpicki *et al* 2002) and the temperature coefficients or the gradients of temperature dependence of the amide protons have been found to span a wide range, -16 to + 4 ppb/K. For a strongly H-bonded amide this value is more positive than -4.5 ppb /K (Baxter and Williamson 1997). This is because the lengthening of the average H-bond distance will be greater for the intermolecular H-bond, such as those with bulk water, than for the intramolecular H-bonds.

However, if the protein structure is not very rigid, as would be the case for many systems, the chemical shifts would also be influenced by local structural and dynamics changes, and then the temperature dependence of chemical shifts may deviate from linearity. Indeed, in certain situations the amide proton chemical shifts have been seen to be non linearly dependent on temperature, and this has been interpreted to indicate existence of alternative conformations the residues can access (Baxter *et al* 1998; Williamson 2003). Identification of such residues provides a description of the energy landscape of the protein in the native state. The observed curvatures can be theoretically deduced as described in the following paragraphs.

Consider a residue having two conformational states accessible to it i.e., a native state and a higher energy state. Following the discussion in the above paragraphs, each of them can be assumed to have a linear variation of chemical shift with temperature as, $\delta_1 = \delta^0_1 + g_1 T$ and $\delta_2 = \delta^0_2 + g_2 T$, where g_1 and g_2 are the gradients of temperature dependence, δ_1 and δ_2 are the chemical shifts of the native and the excited states respectively, and T is the temperature. If P_1 and P_2 are the corresponding populations of the native and the excited states, the observed chemical shift, δ_{obs}, of the amide proton will be given by,

$$\delta_{obs} = \delta_1 P_1 + \delta_2 P_2 \tag{1}$$

These populations depend on the free energy difference between the two states. If there are more states contributing, then the observed shift will be a weighted average over all the accessible states. It is this complex dependence of chemical shifts on many thermodynamic and other factors, which leads to non linear dependence of chemical shifts on temperature. To understand the influence of these factors, simulations of H^N chemical shift variation are performed with temperature in the range, 290 K – 315 K, using a two state model (Krishna Mohan *et al* 2008; Williamson 2003)

$$\delta_{obs} = \frac{(\delta^0_1 + g_1 T) + [(\delta^0_2 + g_2 T)e^{-(\Delta G/RT)}]}{1 + e^{-(\Delta G/RT)}} \tag{2}$$

where, ΔG is the free-energy difference between the two states, and $\Delta G = \Delta H - T\Delta S$, where, ΔH and ΔS are the enthalpy difference and the entropy difference respectively. The results of the simulations are shown in **Fig. 2**. **Fig. 2A** shows the curves for ΔG ranging from 1 - 4 kcal/mol keeping the gradients and chemical shifts of the native and the excited states constant. Here, it is worthwhile to note that in the chosen temperature range (290 K – 315 K) the curvature almost disappears above $\Delta G = 3$ kcal/mol. **Fig. 2B** and **Fig. 2C** show the dependence of curvature on chemical shift differences and gradient differences respectively, between the native and the excited states, when ΔG is held constant ($\Delta G = 2$ kcal/mol). In **Fig. 2B**, three values of δ_l: 8.5 (reference), 8.1 and 9.0 are considered, keeping the other parameters the same as in **Fig. 2A** for convex shape of curvature. Similarly, in **Fig. 2C**, three combinations of gradients: (g_1, g_2) (ppb/K) = (-2, -7) (reference), (-2, -4) and (-4, -7) are considered keeping the other parameters same as in **Fig. 2A** for convex shape of curvature. From these it is evident that neither the chemical shift difference nor the difference in gradients, by itself changes the curvature to a noticeable extent. A simulation carried out for a combination of changes in 'chemical shift difference' and 'gradient difference' ($\delta_1=8.1$ ppm, $\delta_2=8.0$ ppm, $g_1 = -2$ ppb/K, $g_2 = -4$ ppb/K) keeping the free energy constant ($\Delta G = 2$ kcal/mol). This is shown by solid grey line in **Fig 2D**. Interestingly, this curve almost exactly overlaps with the curve for which $\Delta G = 3$ kcal/mol in **Fig. 2A** which has a lower curvature compared to that of the curve with $\Delta G = 2$ kcal/mol; for ease of comparison, the corresponding curve from **Fig. 2A** is redrawn in **Fig. 2D** as a dotted line. This clearly suggests that although the appearance of curvature confirms the presence of alternative states, the lack of curvature does not necessarily imply the absence of low energy excited states. These theoretical simulations will be of great help for interpreting the experimental results on temperature dependence of amide proton chemical shifts.

3. Investigations on native state ruggedness of complex protein systems

A deep well, the bottom of which corresponds to the native state would imply high stability of the native state **(Fig. 1)**. In contrast, a potential well with low lying excited states for the native state would be shallow, and this would have significant influence on the dynamics, structural adaptability, or susceptibility of the protein to various functions (Agarwal 2005; Boehr et al 2010; Eisenmesser et al 2005; Feher and Cavanagh 1999; Kitahara et al 2005; Korzhnev et al 2003; Parak 2003a; Piana et al 2002; Tobi and Bahar 2005). Application of small environmental perturbations such as small concentrations of chemical denaturants, change in pressure, pH change etc., is often useful to investigate the preferential sensitivities of different residues to external perturbations, while the protein itself remains entirely in the native state ensemble (Akasaka 2006; Baxter et al 1998; Chatterjee et al 2007; Kumar et al 2007; Mohan et al 2006; Piana et al 2002). In fact, these environment sensitive residues of polypeptide chains adopt unique 3D structures, they access various near native states which are structurally similar and energetically closer to the native state. These low populated alternative conformations dictate the ruggedness of the native structure and its biological function.

3.1 Differential native state conformational fluctuations in calcium sensor proteins

Calcium ion plays a crucial role in the regulation of various biological processes. To perform several of its functional activities, Ca^{2+} binds to different protein molecules, which are called as calcium binding proteins (CaBPs) (Heizmann and Schafer 1990). CaBPs with EF-hand motif (EF-CaBPs) belong to a growing sub family of CaBPs (Ababou and Desjarlais 2001; Bhattacharya et al 2004; Heizmann 1992; Nelson and Chazin 1998). The EF-hand motif represents the canonical Ca^{2+} binding motif that consists of a contiguous 12 amino-acid residue long loop flanked by two helices (Kretsinger and Nockolds 1973; Strynadka and James 1989). The EF-CaBPs are broadly classified as Ca^{2+} sensors and Ca^{2+} buffers. Ca^{2+} sensors (Finn *et al* 1995; Hanley and Henley 2005; Hilge *et al* 2006; Shaw *et al* 1990; Vinogradova *et al* 2005) such as Calmodulin (CaM), Troponin C (TnC) etc., undergo huge conformational change upon binding to Ca^{2+} whereas, the Ca^{2+} buffers (Hackney *et al* 2005; Lambers *et al* 2006; Rosenbaum *et al* 2006; Vinogradova *et al* 2005) such as Calbindin D$_{9k}$,

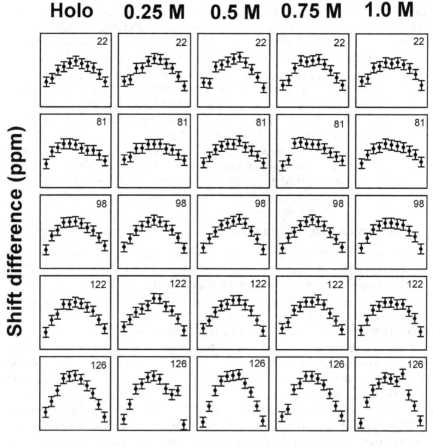

Fig. 3. Illustrative examples for the residues showing nonlinear temperature dependence of backbone ^{1}H^N chemical shifts in *Eh*CaBP as measured in native state and at different concentrations of GdmCl. The measured chemical shifts were fitted to a linear equation. The residuals (observed value – calculated value according to the linear fit) have been plotted against temperature; total scale of y-axis is 0.06 ppm: +0.03 to -0.03 centered at zero, and the temperature range is 280 K – 335 K. The error bars give an indication of the approximate error in measured chemical shifts (±0.004 ppm).

undergo modest conformational changes upon Ca^{2+} binding. In the current section the experimental evidence for the native state ruggedness on a Ca^{2+} sensor protein from the protozoan *Entamoeba histolytica* (*Eh*CaBP), an etiologic agent of amoebiasis has been demonstrated (Atreya *et al* 2001; Bhattacharya *et al* 2006).

As an illustration, **Fig 3** shows the experimentally measured temperature dependence of backbone proton (¹H^N) chemical shifts for few residues in the protein carried out in the temperature range 280 K – 335 K by recording temperature dependent HSQC spectra (Mohan *et al* 2008b). As evident from **Fig. 3**, the observed curvatures are different for different residues. These convex and the concave shapes (Mohan *et al* 2008b; Krishna Mohan *et al* 2008; Williamson 2003) reflect on different kinds of structural perturbations in the excited state compared to the native state as described above in theoretical simulations and illustrated in **Fig 2**. A summary of all the non-linear temperature dependences observed in *Eh*CaBP at different concentrations of GdmCl is given in **Fig. 4**.

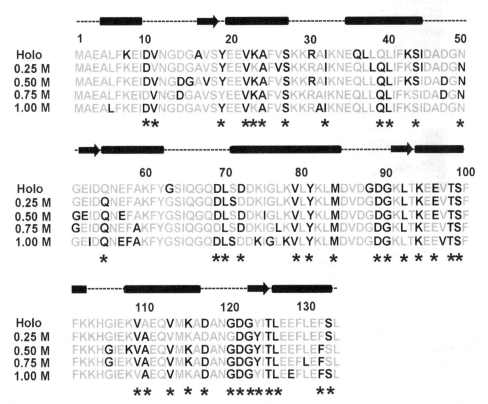

Fig. 4. Residues showing nonlinear temperature dependence of amide proton chemical shifts (black) in the native protein and at different concentrations of GdmCl were shown on the primary sequence of the polypeptide chain. The native secondary structures are shown by arrows (β strands) and cylinders (α helix). Residues accessing alternative conformations at least 3 out of 5 measured GdmCl concentrations are marked with asterisks along the polypeptide chain.

The number of residues accessing alternative states at different concentrations of GdmCl in *Eh*CaBP is (0 M (holo) – 41; 0.25 M – 35; 0.5 M – 50; 0.75 M – 39 and 1.0 M – 42) **(Fig 4)**. The total number of residues which access alternative conformations at least 3 out of 5 measured GdmCl concentrations turned out to be 39 (residues shown with asterisks in Fig. 3), implying that ~ 30 % (39 out of 134) of the residues are accessing alternative conformations. The theoretical simulations described above suggest that the observed curvatures are ~ 2-3 kcal/mol. All these residues are shown in **Fig. 5** on the 3D structure of the protein. Further, the extent of curvatures of individual residues increases or decreases with change in concentration of GdmCl **(Fig.3)** (Williamson 2003). The residues that become more curved with the increasing concentrations of GdmCl are most likely due to the presence of alternate states which are more similar in energy, or more different in shift or gradient to the corresponding native state. This is primarily an off-shoot of contracted unfolding energy landscape in the presence of GdmCl since the GdmCl is not expected to change the nature of the alternative state. Whereas the decrease of curvature can be explained by considering more than one alternative states within 5 kcal/mol, or have an alternative state that becomes very close to native state as GdmCl is increased.

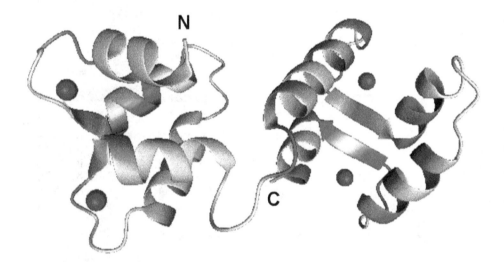

Fig. 5. Residues exhibiting curved temperature dependence at least three out of five concentrations of GdmCl measured in *Eh*CaBP are marked with red color on the 3D structure of the protein (PDB Id: 1JFK).

It is interesting to note that the low energy excited states detected in *Eh*CaBP are not uniformly distributed along the polypeptide chain; different segments of the protein have their own intrinsic preferences to access the alternative conformations. Out of the 39 residues which access low energy excited states, 7 residues belong to EF-hand I; 5 to EF-hand II; 11 to EF-hand III; 13 to EF-hand IV and 3 to the interconnecting loops **(Fig.4)**. It is evident from the data that the density of the conformational fluctuations in the C-terminal domain (24 residues) are twofold compared to the N-terminal counterpart (12 residues)

(**Fig. 4**). This suggests that the C-terminal domain is more flexible and susceptible to structural rearrangements. Further some novel features have been observed in the locations of alternative conformations. The residue at the 5th position of the calcium binding loop (Asp/Asn) that coordinates with Ca^{2+} shows alternative conformations consistently in all the EF-hands. The Gly-6 (the residue at 6th position in the calcium binding loop), which acts as a hinge in the calcium binding loop accesses alternative states in both EF III (Gly-90) and EF IV (Gly-122) hands. Among the calcium binding loops, the IV EF-loop is found to be the most dynamic with maximum number of residues (6 residues out of 12 residue loop) access low energy excited states. This loop has relatively low affinity towards Ca^{2+} compared to the other three loops as demonstrated earlier by the EGTA titration (Mukherjee *et al* 2005), though highly specific for Ca^{2+}. Moreover, recently it has been observed that IV EF-loop also differs with the remaining three EF-loops in the case of Mg^{2+} binding as evidenced by the Mn^{2+} titration (Mukherjee *et al* 2007a). Thus from all the discussion, it can be established that the native state of *EhCaBP* is rugged due to accessing of various alternative states and the ruggedness is more in the C-terminal domain compared to that in the N-terminal domain. It is interesting to note that among the four EF hands, EF-hands I and II belonging to the N-terminal domain show different conformational dynamics from that of EF-hands III and IV belonging to the C-terminal domain (Mohan *et al* 2008b; Mukherjee *et al* 2007b).

Recently Chandra et al (Chandra *et al* 2011) measuring nonlinear temperature dependence of the backbone amide proton chemical shifts on non-myristoylated (non-myr) and myristoylated (myr) neuronal calcium sensor-1 (NCS-1). The authors reported that ~20% of the residues in the protein access alternative conformations in non-myr case, which increases to ~28% for myr NCS-1. These residues are spread over the entire polypeptide stretch and include the edges of α-helices and β-strands, flexible loop regions, and the Ca^{2+}-binding loops. Besides, residues responsible for the absence of Ca–myristoyl switch are also found accessing alternative states. The C-terminal domain is more populated with these residues compared to its N-terminal counterpart. Individual EF-hands in NCS-1 differ significantly in number of alternate states. Such differences in the conformational dynamics between the two domains and among the EF-loops have significant influence on the specificity and affinity of the metal binding properties and also have implications to domain dependent calcium signaling pathways of calcium sensor proteins (Mohan *et al* 2008b; Mukherjee *et al* 2007b).

3.2 Near native states and structure adaptability of dynein light chain protein

Dynein light chain protein (DLC8), a 10.3 kDa protein (89 residues) is the smallest subunit of the Dynein motor complex. DLC8 is a dimer at physiological pH and a stable monomer below pH 4.0 (Barbar and Hare 2004; Mohan *et al* 2006; Nyarko *et al* 2005). The differences between the monomeric and dimeric structures are, (i) the β3 strand in the dimer loses its secondary structure on dissociation to the monomer, and (ii) the helices α1 and α2 and the strands β1 and β2 get shortened by two residues (Liang *et al* 1999; Makokha *et al* 2004). DLC8 dimer acts as a cargo adapter and recognized as interactive "protein hub" (Barbar 2008). The dimer binds the target molecules in an anti-parallel β-strand fashion through its β3-strand, whereas the monomer form of DLC8 is not capable of binding to target proteins

(Alonso *et al* 2001; Fan *et al* 2001; Fan *et al* 1998; Fuhrmann *et al* 2002; Jaffrey and Snyder 1996; Lo *et al* 2001; Naisbitt *et al* 2000; Puthalakath *et al* 1999). This property is expected to have a regulatory role in the protein function.

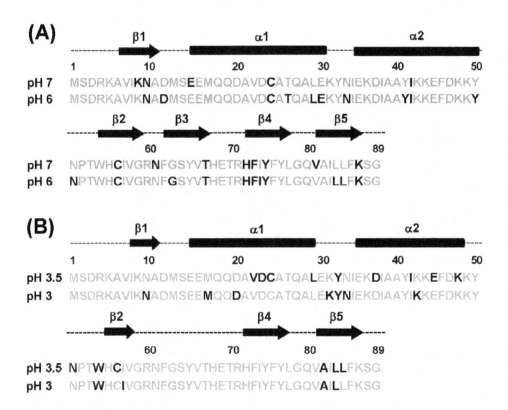

Fig. 6. Residues showing non linear temperature dependence of amide proton chemical shifts (black) in the temperature range is 290 K – 315 K along the polypeptide chain. The results are shown for pH 7.0 and 6.0 (A) and for pH 3.5 and 3.0 (B). The arrows (β strands) and cylinders (α helix) indicate native secondary structures.

Temperature dependence of the amide proton chemical shifts in the DLC8 dimer (pH 7) and in the monomer (pH 3) has been measured in the temperature range 290-315 K (Krishna Mohan *et al* 2008). Among the above mentioned environment perturbations, pH variation is a mild perturbation and in general it changes the protonation states of the various residues depending on the chosen pH range. In order to identify the pH sensitive conformational dynamics in DLC8 protein the temperature dependence of the amide proton chemical shifts in both the dimer and the monomer were measured at slightly different pH conditions i.e., dimer at pH 6 and monomer at pH 3.5. A summary of all these results at various pH values i.e., pH 7 and 6 for the dimer and pH 3.5 and 3 for the monomer are shown in **Fig. 6.** Comparison of **Figs. 6A** and **6B** reveals that the residues accessing alternative conformations have many differences between the dimer and the monomer. The number of residues accessing alternative conformations in the dimer at pH 7 and 6 are 13 and 21 respectively (Fig.6A). Figs. 7A and 7B display the locations of these residues on the native structures of the protein. Likewise, the number of residues accessing alternative conformations in the monomer at pH 3.5 is 15 and that at pH 3 is 11 (Fig.6B). The locations of these residues are marked with red color on the native structure of the monomeric protein in Figs. 7C and 7D respectively.

The differences observed in the positions of the residues accessing alternative conformations in the dimer and in the monomer due to small pH perturbations provide insights into the sensitivity of the conformational fluctuations due to environment perturbations in the two cases. In fact, the perturbation of the dimer landscape would have functional significance since small pH differences are known to exist in different parts of a cell (Spitzer and Poolman 2005; Stewart *et al* 1999; Swietach and Vaughan-Jones 2004; Swietach *et al* 2005; Vaughan-Jones *et al* 2002; Willoughby and Schwiening 2002; Zaniboni *et al* 2003). It is evident from Figs. 7A and 7B that several of the residues that access low energy excited states are surrounding the dimer interface of the molecule which is also the cargo binding site (Krishna Mohan *et al* 2008; Liang *et al* 1999). It can be envisaged that the observed sensitivity of conformational dynamics at the dimer interface due to small environmental perturbations can significantly influence the cargo binding nature of the protein. Likewise, in the monomer (Figs. 7C and 7D) (Krishna Mohan *et al* 2008; Liang *et al* 1999), noticeable differences have been observed in both α1 and α2 helices. Interestingly, the α2 helix participates in several inter-monomer contacts once the dimer is formed and hence its sensitivity to small perturbations may have a crucial role for the proper formation/folding of the functional dimer.

3.2.1 Relationship between sequence, structure and pH sensitivity of DLC8 landscapes

The roughness of the energy landscape and the consequent fluctuations in the native state of a protein is a reflection on the nature of the interactions between the side chains of the different amino acid residues in the three dimensional structure of the protein. While this is not generally predictable, some insights may be obtained in some cases by closely examining the structure and the properties of the amino acids along the sequence. For example, the behaviors of residues with titratable groups, which are likely to be affected by a pH perturbation, can provide useful clues. An observed perturbation at such locations would indicate that conformational fluctuations could be arising due to existence of species with different protonation states; a change in the protonation state of a side–chain causes a local change in the electrostatic potential, and thereby results in some population of an alternative conformation on energetic considerations. Inter-conversion between the major population and the minor population so created leads to the so-called conformational fluctuations.

In the above background it is interesting to note that most of the residues with titratable groups in the side chains in DLC8 (Fig.7E) are located in the regions which are exhibiting conformational fluctuations, and hence, their perturbation by small pH changes provides useful mechanistic insights. In the case of the dimer, the sensitivity of conformational

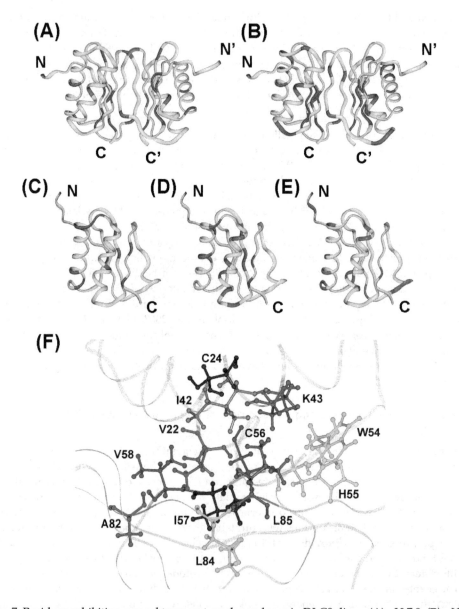

Fig. 7. Residues exhibiting curved temperature dependence in DLC8 dimer **(A)** pH 7.0, **(B)** pH 6.0 (PDB Id: 1f3c) and in monomer **(C)** pH 3.0, **(D)** pH 3.5 (PDB Id: 1hrw), are coloured red on the three dimensional structure of the protein. **(E)** Positions of all the titratable groups in the pH range 7.0 to 3.0 (Aspartates, Glutamates and Histidines) are marked with pink color on the monomer structure (PDB Id: 1rw). **(F)** Zooming in on a particular region surrounding β2 strand in the NMR structure of the monomer (PDB Id: 1rhw) to show the side chain interactions. Only a few residues in α1, α2 and β5 are shown for the sake of clarity.

dynamics can be readily traced to partial protonation of the His side chains as described earlier (Mohan *et al* 2006; Nyarko *et al* 2005). There will be inter-conversions between charged and neutral His and there will also be charge-charge repulsions. These will cause fluctuations in local electrostatic potentials and consequently in local side chain packing, which in turn will affect the main chain conformations. Among the three histidines, His 55 (pK 4.5,), His 68 and His 72 (both have pK of 6.0, (Mohan *et al* 2006; Nyarko *et al* 2005), the latter two would be the major contributors to the observed differences in the fluctuations of the native state in the pH range of 6-7.

In the case of DLC8 monomer His 68 and His 72 do not have any effect on the observed differences as they are completely protonated below pH 4.0. On the other hand His 55 (pK 4.5, (Mohan *et al* 2006; Nyarko *et al* 2005)), would have a significant effect. At pH 3.5 the side chain of His 55 will be protonated to the extent of 90 % and exchange between protonated and free His will contribute to a local dynamics. The environmental perturbation due to this dynamics would get relayed through the β2 strand and the α1, α2 helices and the β5 strand due to the close packing of the side chains in the protein structure **(Fig. 7F)**. The side chains for a few residues of α1, α2 and β5 are shown in the figure and all of these residues are seen to exhibit curved temperature dependence. At pH 3.0, the population of protonated His will increase and this results in the observed perturbation differences. Similarly, the perturbations at the other titratable groups such as aspartates and glutamates in the α1 and α2 helices **(see Fig. 7E)** would also cause local relays and contribute to the accessibility of different low energy excited states. All these influence the native energy landscape of the protein.

3.3 Conformational fluctuations at the phosphorylation site of dynein light chain protein

Recent studies on p21-activated kinase 1 (Pak1), revealed DLC8 as its physiological interacting substrate (binding sites aa 61-89) and the phosphorylation site at Ser 88 (Vadlamudi *et al* 2004). Pak1 phosphorylation of DLC8 on Ser 88 controls vesicle formation and trafficking functions, whereas mutation of Ser 88 to Ala (S88A) prevents macropinocytosis (Song *et al* 2008; Song *et al* 2007; Vadlamudi *et al* 2004; Yang *et al* 2005). Further, DLC8 phosphorylation by Pak1 prevents the interaction with apoptotic protein Bim and plays an essential role in cell survival (Vadlamudi *et al* 2004) and also promotes the dissociation from Intermediate chain (IC74) and hence regulates the assembly of the motor complex (Song *et al* 2007). All these results highlight any perturbation at or near the interface is likely to affect the biological function. Intuitively, the remote effects of any perturbation in a protein must be a consequence of a strong network of interactions which may cause rapid relay of perturbations from any one particular site on the protein structure. However, specific knowledge of how the perturbations travel will be essential in each case to understand the specificities of interactions. In general, perturbations are often introduced deliberately in the form of specific mutations in an attempt to understand the regulatory roles of specific residues involved in target recognition, structural architecture, stability, aggregation and folding features of the wild type protein (Buck *et al* 2007; Frankel *et al* 2007; Grant *et al* 2007; Ishibashi *et al* 2007; Piana *et al* 2008; Riley *et al* 2007; Stollar *et al* 2003).

The phosphorylation site Ser 88 represents an unusual behavior. To understand the conformational behavior phosphorylation site, the amide proton temperature coefficients of

Ser 88 in the WT dimer with those of Ala 88 in the S88A mutant and of Ser 88 in the DLC8 monomer (at pH 3) are measured (Mohan and Hosur 2008). The plots of temperature dependence for these residues are shown in **Fig. 8A**. The measured temperature coefficients are -19.8 ± 0.3, -9.3 ± 0.2 and -5.6 ± 0.1 ppb/°C for Ser 88 in WT dimer, Ala 88 in S88A mutant and Ser 88 in DLC8 monomer respectively. In general the temperature coefficients range between (-2 to -4) ppb/K for a strongly H-bonded amide protons and between (-5 to -10) ppb/K for an exposed solvent accessible (random coil) amide protons (Baxter and Williamson 1997). The values obtained in the case of S88A mutant and DLC8 monomer clearly indicate that the aa 88 is solvent exposed and not strongly hydrogen bonded. On the other hand, a large value of -19.8 ± 0.3 ppb/°C suggests that the environment of Ser 88 in WT dimer is highly susceptible to perturbation. None of the other residues, either in WT monomer or in S88A mutant, showed such a huge temperature coefficient value (Baxter and Williamson 1997; Mohan and Hosur 2008). Williamson et al reported such a large value of temperature coefficient in the herpes simplex virus glycoprotein D-1 antigenic domain (Williamson *et al* 1986) and the experiments demonstrated by Andersen et al (Andersen *et al* 1992) suggested that this amide proton was in fact involved in a transient hydrogen-bonded structure, and thus the large temperature coefficient could be attributed to a loss of secondary/tertiary structure on heating. If the large temperature coefficient of Ser 88 is a consequence of transient hydrogen-bonding and due to loss of secondary/tertiary structure, then Ser 88 should exhibit conformational fluctuations (alternative states). The presence of alternative states for Ser 88 has been tested on Ala 88 residue of S88A mutant and Ser 88 in DLC8 monomer and dimer **(Fig. 8B)**. It is evident that Ser 88 in WT dimer does show a curved temperature dependence of amide proton chemical shifts. Thus, it can be concluded that the amide proton in Ser 88 in WT DLC8 dimer is transiently H-bonded; that it is not a stable H-bond was independently inferred from deuterium exchange studies (Mohan and Hosur 2008; Mohan *et al* 2008a). Moreover, the dimeric structure suggests that the amide group of Gly 89 is very close (~ 2.6 Å) to the backbone nitrogen of Ser 88 suggesting a possibility of potential transient H-bonding.

The mechanism for the relay of perturbations from the Ser 88 site can be envisaged by understanding the close side chain packing. The side chain packing of perturbed residues and Ser 88 are shown in **Fig. 8C**. The crystal structure shows that Ser 88 OG atom is packed against the imidazole ring of His 55 and in addition forms a hydrogen bond with the backbone carbonyl of Ser 88' (Ser 88 of other monomer) (Liang *et al* 1999). From **Fig. 8C** it is evident that Ser 88 is buried inside and packed over side chains of crucial residues at the dimer interface. A closer look at Ser 88 environment in **Fig. 8C** depicts that Ser 88 is closely packed against the side chains of Thr 67, His 68 and Glu 69 of both the monomers in the dimer. Furthermore, the distance measurements between the back bone and side chain atoms of Ser 88 and those of Thr 67, His 68, Glu 69 and Thr 70 indicated that several atoms of Glu 69 are very close (~ 2 - 4 Å), whereas, for residues Thr 67, His 68, Thr 70 there is at least one atom in the distance range of 4 -6 Å. All of these residues are perturbed by the S88A mutation as seen from the chemical shift data. The other perturbed residues are slightly farther (> 6 Å). All these are shown in a color coded manner in **Fig. 8C**. This qualitative analysis provides a mechanistic insight into the relay of perturbation from the phopshorylation site; Glu 69 is most easily perturbed and the disturbance then runs on both sides at the dimer interface. Then, from Tyr 65 the relay spreads to Lys 44 which is engaged in a side chain H-bond.

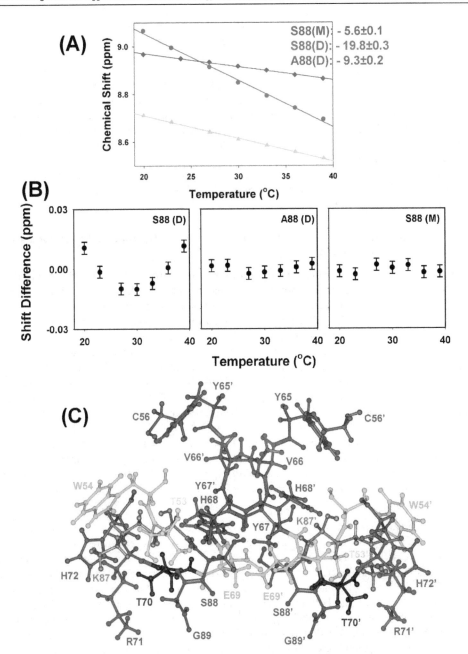

Fig. 8. (A) Graph depicting the temperature dependence of amide proton chemical shifts for Ser 88 in DLC8 WT-dimer (Circles, Red), Ala 88 in S88A mutant at pH 7 (Triangles, Green), and Ser 88 in DLC8 monomer at pH 3 (Diamonds, Blue). The solid lines line represents the best linear fit. (B) Comparison of non linear/linear temperature dependence of amide

proton chemical shifts at amino acid position 88 in DLC8 protein: Ser 88 in DLC8 WT-dimer [S88-(D)], Ala 88 in S88A mutant [A88-(D)] at pH 7 and Ser 88 in DLC8 monomer [S88-(M)] at pH 3. The measured chemical shifts were fitted to a linear equation. The residuals (observed value – calculated value according to the linear fit) have been plotted against temperature; total scale of y-axis is 0.06 ppm: +0.03 to -0.03 centered at zero, and the temperature range is 18 ºC – 40 ºC. **(C)** Zooming in on a particular region of the dimer interface around Ser 88 in the crystal structure (PDB Id: 1cmi), to show accessibility of the phosphorylation site (side chain –OH of Ser 88) and the interactions of side chains various other residues with Ser 88. Side chains of the residues which are perturbed due to S88A are only shown. The different residues are color coded to indicate the proximity of the side chain atoms of the residue to backbone NH of ser 88; Blue: at least one atom of the side chain is within 2-4 Å, Green: at least one atom of the side chain is within the range 4-6 Å, Red: all atoms are beyond 6 Å.

3.4 Alternative conformations in small globular proteins: Sources of fluctuations and implications to function/folding

Experiments have been performed by various research groups to detect the alternative conformations on different monomeric proteins in order to throw light either on the functional implications or on the folding trajectories. Investigations by Kumar et al (Kumar *et al* 2007) on SUMO-1 suggested that the alternative conformations span the length of the protein chain but are located at particular regions on the protein structure. The authors observed that several of the regions of the protein structure that exhibit such fluctuations coincide with the protein's binding surfaces with different substrate like GTPase effector domain (GED) of dynamin, SUMO binding motifs (SBM), E1 (activating enzyme, SAE1/SAE2) and E2 (conjugating enzyme, UBC9) enzymes of sumoylation machinery and speculated that these conformational fluctuations have significant implications for the binding of diversity of targets by SUMO-1. Another report by Srivastava et al (Srivastava and Chary 2011) on hahellin (a βγ-crystallin domain) in its Ca^{2+}-bound form depicted a large conformational heterogeneity with nearly 40% of the residues, some of which are part of Ca2+-binding loops. Further, they observed that out of the two Greek key motifs, the second Greek key motif is floppy as compared to its counterpart.

Extensive research investigations on theoretical and experimental aspects of different protein systems regarding low-energy excited states have been performed by Williamson and co-workers (Baxter *et al* 1998; Tunnicliffe *et al* 2005; Williamson 2003). Studies on conformational ensemble of cytochrome *c* revealed high structural entropy (Williamson 2003). The density of alternative states is particularly high near the heme ligand Met80, which is of interest because both redox change and the first identified stage in unfolding are associated with change in Met80 ligation. By combining theoretical and experimental approaches, it is concluded that the alternative states each comprise approximately five residues, have in general less structure than the native state and are therefore locally unfolded structures. The locations of the alternative states the global unfolding pathway of cytochrome *c*, hinted that they may determine the pathway. Similar experiments on B1 domains of streptococcal proteins G and L (Tunnicliffe *et al* 2005), which are structurally similar, but have different sequences and folding established that several of the residues have curved amide proton temperature and indicated approximately 4–6 local minima for each protein. Further, reports on N-terminal domain of phosphoglycerate kinase, hen egg-white lysozyme, SUMO1 and BPTI (Baxter *et al* 1998; Kumar *et al* 2007) established that conformational heterogeneity arises from a number of independent sources such as, aromatic ring current effects, a minor conformer generated through disulphide bond isomerisation; an alternative hydrogen bond network associated with buried water molecules; alternative hydrogen bonds involving backbone amides and surface-exposed side-chain hydrogen bond acceptors; and the disruption of loops, ends of secondary structural elements and chain termini.

In conclusion, on one hand the ruggedness of the native energy landscape of the protein systems provide rationales for the adaptability of the protein structure to bind various target molecules in order to carry out the biological functions efficient manner. On the other hand, it throws light on many potential unfolding initiation sites in the protein. Furthermore, the origins of these conformational fluctuations provide mechanistic insights into the protein network of hydrophobic/H-bond interactions that dictate the protein stability.

4. Abbreviations

NMR - Nuclear magnetic resonance spectroscopy; HSQC - Hetero nuclear single quantum correlation spectroscopy; DLC8 - Dynein light chain protein; *EhCaBP* - *Entamoeba histolytica* calcium binding protein; Gdmcl - Guanidine Hydrochloride; NCS - Neuronal calcium sensor protein; Myr - Myristoylated; Non-myr - Non-Myristoylated; Pak-1 - P21 activated kinease; SUMO - Small Ubiquitin-like Modifier; BPTI - Bovine pancreatic trypsin inhibitor; UBC9 - Ubiquitin carrier protein 9 .

5. Acknowledgements

The author thank Prof. Ramakrishna V Hosur (TIFR, Mumbai), Prof. K. V. R. Chary (TIFR, Mumbai) for their invaluable guidance, Dr. Sulakshana Mukherjee (UCSD, California) for critical suggestions, the NMR facility at Tata Institute of Fundamental Research (Mumbai, India) and the library facilities at Rutgers University (Piscataway, New Jersey) are greatly acknowledged.

6. References

Ababou A and Desjarlais J R 2001 Solvation energetics and conformational change in EF-hand proteins; *Protein Sci.* 10 301-312.

Agarwal P K 2005 Role of protein dynamics in reaction rate enhancement by enzymes; *J. Am. Chem. Soc.* 127 15248-15256.

Akasaka K 2006 Probing conformational fluctuation of proteins by pressure perturbation; *Chem Rev.* 106 1814-1835.

Alonso C, Miskin J, Hernaez B, Fernandez-Zapatero P, Soto L, Canto C, Rodriguez-Crespo I, Dixon L and Escribano J M 2001 African swine fever virus protein p54 interacts with the microtubular motor complex through direct binding to light-chain dynein; *J. Virol.* 75 9819-9827.

Andersen N H, Chen C P, Marschner T M, Krystek S R, Jr. and Bassolino D A 1992 Conformational isomerism of endothelin in acidic aqueous media: a quantitative NOESY analysis; *Biochemistry* 31 1280-1295.

Anderson N H, Neidigh J W, Harris S M, Lee G M, Liu Z and Tong H 1997 Extracting information from the temperature gradients of polypeptide HN chemical shifts. 1. The importance of conformational averaging; *J. Am. Chem. Soc.* 119 8547-8561.

Atreya H S, Sahu S C, Bhattacharya A, Chary K V and Govil G 2001 NMR derived solution structure of an EF-hand calcium-binding protein from Entamoeba Histolytica; *Biochemistry* 40 14392-14403.

Bai Y, Sosnick T R, Mayne L and Englander S W 1995 Protein folding intermediates: native-state hydrogen exchange; *Science* 269 192-197.

Barbar E 2008 Dynein light chain LC8 is a dimerization hub essential in diverse protein networks; *Biochemistry* 47 503-508.

Barbar E and Hare M 2004 Characterization of the cargo attachment complex of cytoplasmic dynein using NMR and mass spectrometry; *Methods Enzymol.* 380 219-241.

Baxter N J, Hosszu L L, Waltho J P and Williamson M P 1998 Characterisation of low free-energy excited states of folded proteins; *J. Mol. Biol.* 284 1625-1639.

Baxter N J and Williamson M P 1997 Temperature dependence of 1H chemical shifts in proteins; *J. Biomol. NMR* 9 359-369.

Bhabha G, Lee J, Ekiert D C, Gam J, Wilson I A, Dyson H J, Benkovic S J and Wright P E 2011 A dynamic knockout reveals that conformational fluctuations influence the chemical step of enzyme catalysis; *Science* 332 234-238.

Bhattacharya A, Padhan N, Jain R and Bhattacharya S 2006 Calcium-binding proteins of Entamoeba histolytica; *Arch. Med. Res.* 37 221-225.

Bhattacharya S, Bunick C G and Chazin W J 2004 Target selectivity in EF-hand calcium binding proteins; *Biochim. Biophys. Acta* 1742 69-79.

Boehr D D, McElheny D, Dyson H J and Wright P E 2006 The dynamic energy landscape of dihydrofolate reductase catalysis; *Science* 313 1638-1642.

Boehr D D, McElheny D, Dyson H J and Wright P E 2010 Millisecond timescale fluctuations in dihydrofolate reductase are exquisitely sensitive to the bound ligands; *Proc. Natl. Acad. Sci. U. S. A* 107 1373-1378.

Bryngelson J D, Onuchic J N, Socci N D and Wolynes P G 1995 Funnels, pathways, and the energy landscape of protein folding: a synthesis; *Proteins* 21 167-195.

Buck T M, Wagner J, Grund S and Skach W R 2007 A novel tripartite motif involved in aquaporin topogenesis, monomer folding and tetramerization; *Nat. Struct. Mol. Biol.* 14 762-769.

Chandra K, Sharma Y and Chary K V 2011 Characterization of low-energy excited states in the native state ensemble of non-myristoylated and myristoylated neuronal calcium sensor-1; *Biochim. Biophys. Acta* 1814 334-344.

Chatterjee A, Krishna Mohan P M, Prabhu A, Ghosh-Roy A and Hosur R V 2007 Equilibrium unfolding of DLC8 monomer by urea and guanidine hydrochloride: Distinctive global and residue level features; *Biochimie* 89 117-134.

Cierpicki T and Otlewski J 2001 Amide proton temperature coefficients as hydrogen bond indicators in proteins; *J. Biomol. NMR* 21 249-261.

Cierpicki T, Zhukov I, Byrd R A and Otlewski J 2002 Hydrogen bonds in human ubiquitin reflected in temperature coefficients of amide protons; *J. Magn Reson.* 157 178-180.

Clore G M 2011 Exploring sparsely populated states of macromolecules by diamagnetic and paramagnetic NMR relaxation; *Protein Sci.* 20 229-246.

Clore G M and Iwahara J 2009 Theory, practice, and applications of paramagnetic relaxation enhancement for the characterization of transient low-population states of biological macromolecules and their complexes; *Chem. Rev.* 109 4108-4139.

Dill K A and Chan H S 1997 From Levinthal to pathways to funnels; *Nat. Struct. Biol.* 4 10-19.

Dobson C M 2003 Protein folding and misfolding; *Nature* 426 884-890.

Dobson C M and Karplus M 1999 The fundamentals of protein folding: bringing together theory and experiment; *Curr. Opin. Struct. Biol.* 9 92-101.

Dunker A K, Brown C J, Lawson J D, Iakoucheva L M and Obradovic Z 2002 Intrinsic disorder and protein function; *Biochemistry* 41 6573-6582.

Dyson H J and Wright P E 2005 Elucidation of the protein folding landscape by NMR; *Methods Enzymol.* 394 299-321.

Eisenmesser E Z, Millet O, Labeikovsky W, Korzhnev D M, Wolf-Watz M, Bosco D A, Skalicky J J, Kay L E and Kern D 2005 Intrinsic dynamics of an enzyme underlies catalysis; *Nature* 438 117-121.

Fan J, Zhang Q, Tochio H, Li M and Zhang M 2001 Structural basis of diverse sequence-dependent target recognition by the 8 kDa dynein light chain; *J. Mol. Biol.* 306 97-108.

Fan J S, Zhang Q, Li M, Tochio H, Yamazaki T, Shimizu M and Zhang M 1998 Protein inhibitor of neuronal nitric-oxide synthase, PIN, binds to a 17-amino acid residue fragment of the enzyme; *J. Biol. Chem.* 273 33472-33481.

Feher V A and Cavanagh J 1999 Millisecond-timescale motions contribute to the function of the bacterial response regulator protein Spo0F; *Nature* 400 289-293.

Finn B E, Evenas J, Drakenberg T, Waltho J P, Thulin E and Forsen S 1995 Calcium-induced structural changes and domain autonomy in calmodulin; *Nat. Struct. Biol.* 2 777-783.

Frankel B A, Tong Y, Bentley M L, Fitzgerald M C and McCafferty D G 2007 Mutational analysis of active site residues in the Staphylococcus aureus transpeptidase SrtA; *Biochemistry* 46 7269-7278.

Fraser J S, Clarkson M W, Degnan S C, Erion R, Kern D and Alber T 2009 Hidden alternative structures of proline isomerase essential for catalysis; *Nature* 462 669-673.

Fuhrmann J C, Kins S, Rostaing P, El F O, Kirsch J, Sheng M, Triller A, Betz H and Kneussel M 2002 Gephyrin interacts with Dynein light chains 1 and 2, components of motor protein complexes; *J. Neurosci.* 22 5393-5402.

Grant M A, Lazo N D, Lomakin A, Condron M M, Arai H, Yamin G, Rigby A C and Teplow D B 2007 Familial Alzheimer's disease mutations alter the stability of the amyloid beta-protein monomer folding nucleus; *Proc. Natl. Acad. Sci. U. S. A* 104 16522-16527.

Hackney C M, Mahendrasingam S, Penn A and Fettiplace R 2005 The concentrations of calcium buffering proteins in mammalian cochlear hair cells; *J. Neurosci.* 25 7867-7875.

Hanley J G and Henley J M 2005 PICK1 is a calcium-sensor for NMDA-induced AMPA receptor trafficking; *EMBO J.* 24 3266-3278.

Heizmann C W 1992 Calcium-binding proteins: basic concepts and clinical implications; *Gen. Physiol Biophys.* 11 411-425.

Heizmann C W and Schafer B W 1990 Internal calcium-binding proteins; *Semin. Cell Biol.* 1 277-282.

Hilge M, Aelen J and Vuister G W 2006 Ca2+ regulation in the Na+/Ca2+ exchanger involves two markedly different Ca2+ sensors; *Mol. Cell* 22 15-25.

Ishibashi M, Tatsuda S, Izutsu K, Kumeda K, Arakawa T and Tokunaga M 2007 A single Gly114Arg mutation stabilizes the hexameric subunit assembly and changes the substrate specificity of halo-archaeal nucleoside diphosphate kinase; *FEBS Lett.* 581 4073-4079.

Jaffrey S R and Snyder S H 1996 PIN: an associated protein inhibitor of neuronal nitric oxide synthase; *Science* 274 774-777.

Kitahara R, Yokoyama S and Akasaka K 2005 NMR snapshots of a fluctuating protein structure: ubiquitin at 30 bar-3 kbar; *J. Mol. Biol.* 347 277-285.

Korzhnev D M, Karlsson B G, Orekhov V Y and Billeter M 2003 NMR detection of multiple transitions to low-populated states in azurin; *Protein Sci.* 12 56-65.

Kretsinger R H and Nockolds C E 1973 Carp muscle calcium-binding protein. II. Structure determination and general description; *J. Biol. Chem.* 248 3313-3326.

Krishna Mohan P M, Barve M, Chatterjee A, Ghosh-Roy A and Hosur R V 2008 NMR comparison of the native energy landscapes of DLC8 dimer and monomer; *Biophys. Chem* 134 10-19.

Kumar A, Srivastava S and Hosur R V 2007 NMR characterization of the energy landscape of SUMO-1 in the native-state ensemble; *J. Mol. Biol.* 367 1480-1493.

Lambers T T, Mahieu F, Oancea E, Hoofd L, de L F, Mensenkamp A R, Voets T, Nilius B, Clapham D E, Hoenderop J G and Bindels R J 2006 Calbindin-D28K dynamically controls TRPV5-mediated Ca2+ transport; *EMBO J.* 25 2978-2988.

Levinthal C 1969. Mössbauer Spectroscopy in Biological Systems. (eds. DeBrunner JTP and Munck E), pp 22-24. University of Illinois Press: Illinois.

Liang J, Jaffrey S R, Guo W, Snyder S H and Clardy J 1999 Structure of the PIN/LC8 dimer with a bound peptide; *Nat. Struct. Biol.* 6 735-740.

Lo K W, Naisbitt S, Fan J S, Sheng M and Zhang M 2001 The 8-kDa dynein light chain binds to its targets via a conserved (K/R)XTQT motif; *J. Biol. Chem.* 276 14059-14066.

Makokha M, Huang Y J, Montelione G, Edison A S and Barbar E 2004 The solution structure of the pH-induced monomer of dynein light-chain LC8 from Drosophila; *Protein Sci.* 13 727-734.

Mittermaier A and Kay L E 2006 New tools provide new insights in NMR studies of protein dynamics; *Science* 312 224-228.

Mohan P M, Barve M, Chatterjee A and Hosur R V 2006 pH driven conformational dynamics and dimer-to-monomer transition in DLC8; *Protein Sci.* 15 335-342.

Mohan P M, Chakraborty S and Hosur R V 2008a Residue-wise conformational stability of DLC8 dimer from native-state hydrogen exchange; *Proteins.*

Mohan P M and Hosur R V 2008 NMR characterization of structural and dynamics perturbations due to a single point mutation in Drosophila DLC8 dimer: functional implications; *Biochemistry* 47 6251-6259.

Mohan P M, Mukherjee S and Chary K V 2008b Differential native state ruggedness of the two Ca2+-binding domains in a Ca2+ sensor protein; *Proteins* 70 1147-1153.

Mukherjee S, Kuchroo K and Chary K V 2005 Structural characterization of the apo form of a calcium binding protein from Entamoeba histolytica by hydrogen exchange and its folding to the holo state; *Biochemistry* 44 11636-11645.

Mukherjee S, Mohan P M and Chary K V 2007a Magnesium promotes structural integrity and conformational switching action of a calcium sensor protein; *Biochemistry* 46 3835-3845.

Mukherjee S, Mohan P M, Kuchroo K and Chary K V 2007b Energetics of the native energy landscape of a two-domain calcium sensor protein: distinct folding features of the two domains; *Biochemistry* 46 9911-9919.

Naisbitt S, Valtschanoff J, Allison D W, Sala C, Kim E, Craig A M, Weinberg R J and Sheng M 2000 Interaction of the postsynaptic density-95/guanylate kinase domain-associated protein complex with a light chain of myosin-V and dynein; *J. Neurosci.* 20 4524-4534.

Nelson M R and Chazin W J 1998 Structures of EF-hand Ca(2+)-binding proteins: diversity in the organization, packing and response to Ca2+ binding; *Biometals* 11 297-318.

Nyarko A, Cochrun L, Norwood S, Pursifull N, Voth A and Barbar E 2005 Ionization of His 55 at the dimer interface of dynein light-chain LC8 is coupled to dimer dissociation; *Biochemistry* 44 14248-14255.

Onuchic J N, Luthey-Schulten Z and Wolynes P G 1997 Theory of protein folding: the energy landscape perspective; *Annu. Rev. Phys. Chem.* 48 545-600.

Parak F G 2003a Proteins in action: the physics of structural fluctuations and conformational changes; *Curr. Opin. Struct. Biol.* 13 552-557.

Parak F G 2003b Proteins in action: the physics of structural fluctuations and conformational changes; *Curr. Opin. Struct. Biol.* 13 552-557.

Piana S, Carloni P and Parrinello M 2002 Role of conformational fluctuations in the enzymatic reaction of HIV-1 protease; *J. Mol. Biol.* 319 567-583.

Piana S, Laio A, Marinelli F, Van T M, Bourry D, Ampe C and Martins J C 2008 Predicting the effect of a point mutation on a protein fold: the villin and advillin headpieces and their Pro62Ala mutants; *J. Mol. Biol.* 375 460-470.

Popovych N, Sun S, Ebright R H and Kalodimos C G 2006 Dynamically driven protein allostery; *Nat. Struct. Mol. Biol.* 13 831-838.

Puthalakath H, Huang D C, O'Reilly L A, King S M and Strasser A 1999 The proapoptotic activity of the Bcl-2 family member Bim is regulated by interaction with the dynein motor complex; *Mol. Cell* 3 287-296.

Riley P W, Cheng H, Samuel D, Roder H and Walsh P N 2007 Dimer dissociation and unfolding mechanism of coagulation factor XI apple 4 domain: spectroscopic and mutational analysis; *J. Mol. Biol.* 367 558-573.

Rosenbaum E E, Hardie R C and Colley N J 2006 Calnexin is essential for rhodopsin maturation, Ca2+ regulation, and photoreceptor cell survival; *Neuron* 49 229-241.

Schwarzinger S, Kroon G J, Foss T R, Chung J, Wright P E and Dyson H J 2001 Sequence-dependent correction of random coil NMR chemical shifts; *J. Am. Chem. Soc.* 123 2970-2978.

Schwarzinger S, Kroon G J, Foss T R, Wright P E and Dyson H J 2000 Random coil chemical shifts in acidic 8 M urea: implementation of random coil shift data in NMRView; *J. Biomol. NMR* 18 43-48.

Shaw G S, Hodges R S and Sykes B D 1990 Calcium-induced peptide association to form an intact protein domain: 1H NMR structural evidence; *Science* 249 280-283.

Shoemaker B A, Wang J and Wolynes P G 1999 Exploring structures in protein folding funnels with free energy functionals: the transition state ensemble; *J. Mol. Biol.* 287 675-694.

Song C, Wen W, Rayala S K, Chen M, Ma J, Zhang M and Kumar R 2008 Serine 88 Phosphorylation of the 8-kDa Dynein Light Chain 1 Is a Molecular Switch for Its Dimerization Status and Functions; *J. Biol. Chem* 283 4004-4013.

Song Y, Benison G, Nyarko A, Hays T S and Barbar E 2007 Potential role for phosphorylation in differential regulation of the assembly of dynein light chains; *J. Biol. Chem* 282 17272-17279.

Spitzer J J and Poolman B 2005 Electrochemical structure of the crowded cytoplasm; *Trends Biochem. Sci.* 30 536-541.

Srivastava A K and Chary K V 2011 Conformational heterogeneity and dynamics in a betagamma-Crystallin from Hahella chejuensis; *Biophys. Chem.* 157 7-15.

Stewart A K, Boyd C A and Vaughan-Jones R D 1999 A novel role for carbonic anhydrase: cytoplasmic pH gradient dissipation in mouse small intestinal enterocytes; *J. Physiol* 516 (Pt 1) 209-217.

Stollar E J, Mayor U, Lovell S C, Federici L, Freund S M, Fersht A R and Luisi B F 2003 Crystal structures of engrailed homeodomain mutants: implications for stability and dynamics; *J. Biol. Chem* 278 43699-43708.

Strynadka N C and James M N 1989 Crystal structures of the helix-loop-helix calcium-binding proteins; *Annu. Rev. Biochem.* 58 951-998.

Swietach P, Leem C H, Spitzer K W and Vaughan-Jones R D 2005 Experimental generation and computational modeling of intracellular pH gradients in cardiac myocytes; *Biophys. J.* 88 3018-3037.

Swietach P and Vaughan-Jones R D 2004 Novel method for measuring junctional proton permeation in isolated ventricular myocyte cell pairs; *Am. J. Physiol Heart Circ. Physiol* 287 H2352-H2363.

Tang C, Louis J M, Aniana A, Suh J Y and Clore G M 2008 Visualizing transient events in amino-terminal autoprocessing of HIV-1 protease; *Nature* 455 693-696.

Tilton R F, Jr., Dewan J C and Petsko G A 1992 Effects of temperature on protein structure and dynamics: X-ray crystallographic studies of the protein ribonuclease-A at nine different temperatures from 98 to 320 K; *Biochemistry* 31 2469-2481.

Tobi D and Bahar I 2005 Structural changes involved in protein binding correlate with intrinsic motions of proteins in the unbound state; *Proc. Natl. Acad. Sci. U. S. A* 102 18908-18913.

Tunnicliffe R B, Waby J L, Williams R J and Williamson M P 2005 An experimental investigation of conformational fluctuations in proteins G and L; *Structure. (Camb.)* 13 1677-1684.

Tzeng S R and Kalodimos C G 2009 Dynamic activation of an allosteric regulatory protein; *Nature* 462 368-372.

Uversky V N 2002 Natively unfolded proteins: a point where biology waits for physics; *Protein Sci.* 11 739-756.

Uversky V N 2003 Protein folding revisited. A polypeptide chain at the folding-misfolding-nonfolding cross-roads: which way to go?; *Cell Mol. Life Sci.* 60 1852-1871.

Vadlamudi R K, Bagheri-Yarmand R, Yang Z, Balasenthil S, Nguyen D, Sahin A A, den H P and Kumar R 2004 Dynein light chain 1, a p21-activated kinase 1-interacting substrate, promotes cancerous phenotypes; *Cancer Cell* 5 575-585.

Vaughan-Jones R D, Peercy B E, Keener J P and Spitzer K W 2002 Intrinsic H(+) ion mobility in the rabbit ventricular myocyte; *J. Physiol* 541 139-158.

Villali J and Kern D 2010 Choreographing an enzyme's dance; *Curr. Opin. Chem. Biol.* 14 636-643.

Vinogradova M V, Stone D B, Malanina G G, Karatzaferi C, Cooke R, Mendelson R A and Fletterick R J 2005 Ca(2+)-regulated structural changes in troponin; *Proc. Natl. Acad. Sci. U. S. A* 102 5038-5043.

Whitten S T, Garcia-Moreno E B and Hilser V J 2005 Local conformational fluctuations can modulate the coupling between proton binding and global structural transitions in proteins; *Proc. Natl. Acad. Sci. U. S. A* 102 4282-4287.

Williamson M P 2003 Many residues in cytochrome c populate alternative states under equilibrium conditions; *Proteins* 53 731-739.

Williamson M P, Hall M J and Handa B K 1986 1H-NMR assignment and secondary structure of a herpes simplex virus glycoprotein D-1 antigenic domain; *Eur. J. Biochem.* 158 527-536.

Willoughby D and Schwiening C J 2002 Electrically evoked dendritic pH transients in rat cerebellar Purkinje cells; *J. Physiol* 544 487-499.

Wishart D S, Bigam C G, Holm A, Hodges R S and Sykes B D 1995 1H, 13C and 15N random coil NMR chemical shifts of the common amino acids. I. Investigations of nearest-neighbor effects; *J. Biomol. NMR* 5 67-81.

Wishart D S and Sykes B D 1994 Chemical shifts as a tool for structure determination; *Methods Enzymol.* 239 363-392.

Wolynes P G 2005 Energy landscapes and solved protein-folding problems; *Philos. Transact. A Math. Phys. Eng Sci.* 363 453-464.

Wolynes P G, Onuchic J N and Thirumalai D 1995 Navigating the folding routes; *Science* 267 1619-1620.

Wüthrich K 1986. *NMR of protein and nucleic acids*. John Wiley and Sons: New York.

Yang Z, Vadlamudi R K and Kumar R 2005 Dynein light chain 1 phosphorylation controls macropinocytosis; *J. Biol. Chem* 280 654-659.

Zaniboni M, Swietach P, Rossini A, Yamamoto T, Spitzer K W and Vaughan-Jones R D 2003 Intracellular proton mobility and buffering power in cardiac ventricular myocytes from rat, rabbit, and guinea pig; *Am. J. Physiol Heart Circ. Physiol* 285 H1236-H1246.

4

Conformational and Disorder to Order Transitions in Proteins: Structure / Function Correlation in Apolipoproteins

José Campos-Terán[1], Paola Mendoza-Espinosa[2],
Rolando Castillo[3] and Jaime Mas-Oliva[2]

[1]*Departamento de Procesos y Tecnología, DCNI, Universidad Autónoma Metropolitana,*
[2]*Instituto de Fisiología Celular, Universidad Nacional Autónoma de México,*
[3]*Instituto de Física, Universidad Nacional Autónoma de México*
México, D.F., México

1. Introduction

The concept of protein folding is directly related with the process of reversible disorder-to-order transitions, by which an unfolded polypeptide chain folds into a specific functional native structure (Eaton et al., 2000; Rose et al., 2006). For folding into a native state, unfolded polypeptide chains require the intervention of weak interactions. Driven by hydrophobic interactions, a polypeptide chain begins to fold when placed in an aqueous medium, and rapidly becomes a molten globule followed by an important release of latent heat. Stabilization of the molten globule is achieved mainly through the distribution of hydrophobic residues away from the water matrix. On the other hand, because the polar residues contained in a protein develop hydrogen bonds with the water network as well as with each other, α-helices and β-sheets can be formed when bonds switch between molecules. It has been calculated that such bonds might be in the order of 10^{-12} s, very similar to those we find in water itself. The random equilibrium can be shifted toward one of these conformations by means of two stages: a fast stage, during which the unfolded polypeptide becomes a molten globule; and a slow stage, in which the molten globule slowly transforms into a fully folded form or native state (Huang, 2005). These two stages in protein folding can be illustrated by a "folding funnel", during which due to a small change in entropy with a large loss of energy, a molten globule evolves into the native state (Fig. 1a) (Dobson, 2003; Gsponer & Vendruscolo, 2006).

Although the process is extremely efficient, there is always the possibility that this accurate mechanism might fail, and the possibility of finding a protein folded into a non-native state becomes a reality (Dobson, 1999). Proteins that follow this pathway might present transiently stable conformations, promoting their interaction with other molecules and facilitating the fact that they might form amorphous oligomers and end in a state of aggregation. Aggregation does not arise from a random coil state, but rather from a series of intermediates that—based on the type of secondary structure acquired during folding—

might or might not resemble the native state (Fig. 1b) (D. Eisenberg et al., 2006, Gsponer & Vendruscolo, 2006). It is well known now that primary polypeptide sequences become the key factor during this process, while the environment surrounding the protein is an important factor for explaining the folding process (Fink, 1998). On the other hand, natively unfolded proteins, known to lack the presence of permanent secondary and tertiary structures, have been recognized at least in the absence of other proteins, to present the tendency to organize themselves into amyloidogenic structures. Considering that the native state is located at the lowest minimum of the "folding funnel", it indicates that this region is the most thermodynamically stable configuration of the polypeptide chain under physiological conditions. For proteins, whose functional state is a tightly packed globular fold, a key step in fibril formation related to partial or complete unfolding is less likely to occur and therefore remains protected against aggregation (Dobson, 2004). In this respect, it has been proposed that the more transient structures thus formed in proteins, the better probability for key determinants in amyloid fibril formation to be found (Ohnishi & Takano, 2004). Thus, many of the known forms of amyloid diseases associated with genetic mutations that decrease protein stability and promote unfolding (Ohnishi & Takano, 2004), are both related to disorder-to-order conformational transitions.

The first experimental evidence about a specific disorder-to-order transition was presented over 30 years ago with the mechanism description for the conversion of trypsinogen to trypsin (Bode & Huber, 1976). This mechanism is characterized by the enzymatic removal of a hexapeptide from the N-terminal region of trypsinogen in order to form trypsin. This basic change promotes the transition from a disordered state of the "specificity pocket" in trypsinogen to an ordered state in trypsin (Huber & Bode, 1978). Since it is known that several amino acids that make up a protein strongly favor a disordered state, at present this "new view" of folding is beginning to be further studied, in which the influence of external or environmental conditions sustains well-tested transitions between disordered and ordered states. Specific polypeptide chains contained in proteins or complete proteins

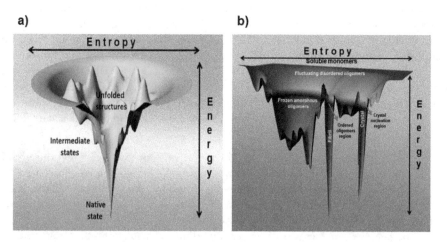

Fig. 1. a) Folding funnel energy landscape b) Protein aggregation energy landscape.

lacking defined tertiary structures are known to have the capacity to undergo disorder-to-order transitions upon binding to specific (Tompa, 2002) or multiple partners (James & Tawfik, 2003). It is precisely this ability that allows the concept of "protein disorder" to be proposed as an important feature in the capability of proteins to present regions with switching properties (Bustos & Iglesias, 2006; Dalal & Regan, 2000; Kriwacki et al., 1996).

From an evolutionary point of view, it appears that intrinsic disorder in proteins might have been the driving force behind many of the adaptability processes found in proteins (Dobson, 1999; Dunker et al., 1998). Taking into account that the number of proteins presenting disordered regions directly related with function and therefore with disease is increasingly growing, an interest to also generate accessible data banks for improving information management has increased. Therefore, the database of disordered proteins (DisProt) was created and released in August 2006 by the group of Dunker (Sickmeier et al., 2007) with extremely good results at present (Cortese et al., 2008). Since then, other systems for studying disorder in proteins have been released, such as the Integrated Protein Disorder Analyzer, which aims at identifying and predicting disordered region in proteins (Su et al., 2007), or algorithms for predicting and evaluating aggregation "hot spots" (AGGRESCAN) (Conchillo-Sole´ et al., 2007). According to Dunker's group and as predicted by the Predictor of Natural Disordered Regions (PONDR) server (Romero et al., 2001), a large percentage of all proteins involved with some sort of a disease have been identified as directly related with disordered regions in proteins closely associated with signaling. From a general point of view, disordered regions in proteins have been divided into the following two classes: the class in which proteins retain a low percentage of secondary structure together with unstable tertiary structures during a molten globule state, recognized as the collapsed class; and second, the extended class in which proteins with a highly extended backbone resemble a β-sheet conformation (Dunker et al., 2001; Uversky, 2002).

In general, proteins containing disordered regions have been recognized as associated with several human diseases, including cardiovascular disease, cancer, degenerative diseases, and diabetes. Interestingly, because in many of these cases cell signaling function has been involved, there is a strong possibility that disorder-to-order transitions in proteins playing normal switching roles in the cell might become distorted and therefore abolish or transform the normal protein–protein language into an aberrant one. Therefore, the basic properties of a switching mechanism must be based on the equilibrium between high specificity and weak affinities accompanied by a large conformational entropy decrease. This phenomenon is based principally on the fact that upon binding, disorder-to-order transitions can overcome steric restrictions and thereby enable larger interaction surfaces in protein–protein complexes than those that could be obtained for rigid partners. Despite the extraordinary importance of this type of transition, we continue to lack detailed biophysical studies that might demonstrate a close relationship between this type of disorder-to-order organization and protein function.

In an attempt to define the possibility that folding key features in proteins could provide us with the manner in which to explain basic issues such as receptor recognition, lipid transfer activity, and self-exchangeability carried out by several lipid transfer proteins including Apolipoproteins (Apos), our group has attempted to address these points by directly measuring molecular conformational changes of Apos at air/water and lipid/water interfaces, in order to approach the possible mechanisms that might explain these phenomena (Xicohtencatl-Cortes et al., 2004a, 2004b). As described below, this has been achieved employing Langmuir monolayers in conjunction with Brewster angle microscopy (BAM), atomic force microscopy (AFM) of Apos LB films (Bolaños-García et al., 1999, 2001; Mas-Oliva et al., 2003; Xicohtencatl-Cortes et al., 2004a), grazing incidence X-ray diffraction on protein monolayers (Ruíz-García et al., 2003), and surface force measurements (SFA) (Campos-Terán et al., 2004; Ramos et al., 2008). Because at that time, we were unable to define whether the secondary structure of specific segments of Apolipoprotein CI (ApoCI) and AII (Apo AII) remained stable independently of their position at air/water and lipid/water interfaces, recently we have addressed the possibility that these segments responding to specific environmental changes and following disorder-to-order transitions might function as molecular switches that trigger function (Mendoza-Espinosa et al., 2008, 2009). Moreover, following the same approach with specific peptides synthesized from the reported structure of Apolipoprotein AI (Apo AI), we have found that when left in water at 4°C a very slow disorder-to-order transition develops over the course of days, from a fully disordered state to a well-developed β-sheet secondary structure. This behavior further supports the fact that the physicochemical characteristics of the environment must be considered as a key factor in the equilibrium displacement within the secondary structure of a protein or specific segments toward α-helices or β-sheets (Andreola et al., 2003). Here, the result that specific segments of Apo AI slowly develop fibril-like structures indicates the possibility that pathological processes such as atherogenesis might be also considered as an amyloidotic-related process (Fig. 2) (Mendoza-Espinosa et al., 2009; Westermark et al., 1995). New results related to these studies are also described in this chapter.

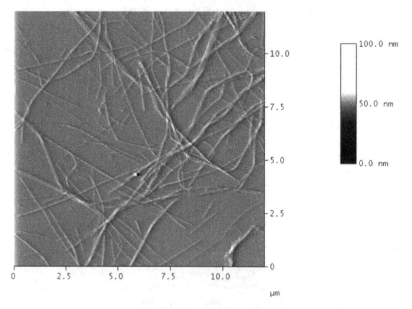

Fig. 2. Atomic force microscopy image (12 x 12 μm) of apolipoprotein AI-peptide DRV fibrils (amino acids 9–24). Fibrils show an average length of 300 nm and 25 nm in height.

2. Structural characteristics of Apolipoproteins CI, AII and AI

Apolipoproteins (Apos) are membrane active proteins that are constituents of high-density lipoproteins (HDL), which are related to the reverse cholesterol transport (Despres et al., 2000). These proteins have an amphiphilic character, since a polar protein face is formed by charged amino acid residues clustered on one side of the α-helices, whereas a hydrophobic surface composed of non-polar residues is formed at the opposite face (Bolaños-García et al., 1997). When Apos are in contact with a polar/non-polar media, their natural tendency is to anchor the hydrophilic and hydrophobic regions in the polar and in the non-polar media, respectively. Thus, a hydrophobic/hydrophilic interface tends to induce a specific orientation on the adsorbed molecules. Some lipoprotein-bound Apos are able to dissociate from the lipoprotein surface in a lipid-poor form, and then transferred through the plasma serum to other lipoproteins (Castro & Fielding, 1984; Clay et al., 1999; Liang et al., 1995; Wang, 2002; Weinberg & Spector, 1985). Although, this mechanism is poorly understood, it is known to be conducted by interactions between Apolipoproteins located at the lipid surface. Apo CI, AII and AI are members of this family of proteins that apparently give lipoproteins directionality and the ability to interact with receptors at the surface of cells.

Apo CI is composed of 57 amino acid residues in length, with a molecular mass of 6.63 KDa. This protein plays a key role in the chylomicron uptake (S. Eisenberg, 1990) and in the regulation of apolipoprotein-E/β-VLDL (very low-density lipoproteins) particle interaction (Swaney & Weisgraber, 1994). Secondary structure predictions, nuclear magnetic resonance, and circular dichroism studies made on Apo CI have revealed a high α-helix content, distributed in two α-helices (Bolaños-García et al., 1999). The first α-helix (residues 4-30) presents approximately 7.5 periods, while the second one (residues 35-53) consists of 5.2 periods (see Fig. 3). In addition both α-helices present important hydrophobic moments (μH) (Bolaños-García et al., 1999). Apo AII is the second major apolipoprotein of high-density lipoproteins (HDL) and it is synthesized in the liver (Eggerman et al., 1991). This protein has been suggested as a modulator of reverse cholesterol transport rather than a strong determinant of lipid metabolism (Tailleux et al., 2002). Apo AII is formed by two identical polypeptide chains connected by a disulfide bridge at position 6, where each chain corresponds to 77 amino acid residues in length and a molecular mass of 8.708 kDa (Brewer et al., 1972, 1986). Predictive and circular dichroism studies (Bolaños-García et al., 1997, 2001), as well as high-resolution crystal structure studies (Kumar et al., 2002) have shown that each chain of the Apo AII presents two α-helix motifs (segments encompassing 7-27 and 32-67) as its main secondary structure (see Fig. 3). These α-helices present an important hydrophobic moment, have approximately 31.5 and 54 Å in length and they are connected by a short peptide chain as a loose hinge (Bolaños-García et al., 2001). Correlation between protein stability to thermal denaturation and secondary structure content has also been investigated (Bolaños-García et al., 2001).

Apo AI has been studied in its free state and membrane models due to the importance that involves understanding the processes that give rise to nascent HDL, as well as the precise mechanisms that support these phenomena in relationship with the process of reverse cholesterol transport. The 243 amino acid polypeptide chain of the Apo AI is organized in blocks of 22 and 11 residues, which are predicted to form helix-type amphipathic

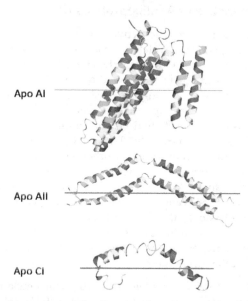

Apo AI

Apo AII

Apo CI

Fig. 3. Secondary structure images of Apolipoproteins showing their α-helical conformation. The color code for the residues is as follows: aromatic-magenta, aliphatic-yellow, polar non charged-green, positively charged-blue, negatively charged-red.

segments (see Fig. 3). The helices that make up the Apo AI, have been classified as follows: (1-45 aa) G*, (44-65,66-87,121-142,143-164,165-186,187-208 aa) A_1, (88-98,99 - 120,209-219,220-241) Y. Helices classified as type G correspond to amphipathic helices that form the interior of globular proteins, reason why amino acids they contain correspond to a hydrophobic type character. Amphipathic helices of the A_1 type have as a characteristic the presence of positively charged amino acids at the hydrophobic/hydrophilic interface, while the negative residues are in the center of the polar face. On the other hand, Y-type helices present the characteristic of having positive charged aminoacids separated by negative ones (Segrest et al., 1992). Currently, there are two crystal structures of the lipid-free Apo AI in different conformations. In the crystal structure obtained by Borhani et al. (Δ1-43) (Borhani et al., 1997), the N-terminal segment is truncated. This structure is unique in presenting a conformation similar to the one that would be in the presence of lipids. Also, the Apo AI (1-243) structure obtained by Ajees et al. (Ajees et al., 2006), presents two domains formed by four α-helices in the N-terminal and 2 α-helices in the C-terminus. Spectroscopic techniques have shown that the lipid-free Apo AI in solution presents a three-dimensional arrangement in two domains similar to that observed in the crystal structure, but with much less organization (Tanaka et al., 2008).

3. Monolayer behavior of Apolipoproteins

When Apolipoproteins are in contact with a polar/non-polar media, they will anchor the hydrophilic and hydrophobic regions in the polar and in the non-polar media, respectively. Thus, a hydrophobic/hydrophilic interface tends to induce a specific orientation on the adsorbed proteins. As mentioned, Apo CI, AII and AI are associated with lipoproteins particles that are modeled (Borhani et al., 1997) as spheres with a shell of a phospholipid monolayer, with the polar head groups oriented towards the aqueous phase, and the core consists of triglycerides and cholesterol esters (hydrophobic region). In these models, Apos are usually oriented parallel to the surface of the lipoprotein particles. A way to understand the behavior of Apos on the lipoprotein surface is to deposit them on an interface that models the lipoprotein surface, which could be increasingly complex as needed.

The first attempts in this direction have used Apo CI and AII Langmuir monolayers deposited at the air/water interface (Bolaños-García et al., 1999, 2001). For both proteins the compression isotherm showed two first-order phase transitions (see Fig. 4). The first one corresponds to the coexistence between a liquid (L) and a gaseous (G) phase where the proteins have low interaction. The second transition involves two condensed phases; the liquid phase, L, and a condensed phase denoted by LC. For the case of Apo CI, this second transition occurs at a surface pressure (Π) of approximately 33 mNm^{-1} and at an area (A) between 350 and 600 Å^2/molecule. For Apo AII it was found at $\Pi \sim$30-35 mNm^{-1} and

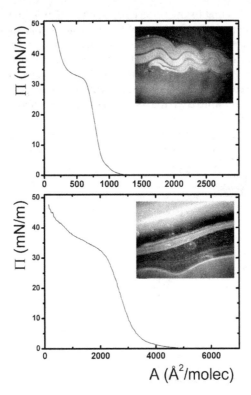

Fig. 4. Langmuir monolayer isotherms of Apo CI (upper panel) and Apo AII (lower panel) at 25.1 ºC. Both proteins were dispersed over a phosphate buffer subphase (pH=8.0) containing 3.5 M KCl. Insets show BAM images at the L-LC coexistence. Adapted from (Xicohtencatl-Cortes et al., 2004a).

A~1000-2500 Å²/molecule. Brewster angle microscopy (BAM) images taken at this transition showed the L phase as dark regions while the LC phase was clearly observed as very bright domains. In the liquid phase, the protein configurations are restricted to a horizontal orientation at the interface due to the amphiphilic character of these proteins. As the surface area is decreased on isothermal compression, one of the α-helix segments for the case of Apo CI and two for Apo AII are expelled from the interface. Direct evidence of this conformational change, as well as of the α-helix structure of Apo CI and AII, have been shown using grazing incidence X-ray diffraction and atomic force microscopy (AFM) of Langmuir-Blodgett (LB) films of transferred protein monolayers (Ruíz-García et al., 2003). It is important to mention that a similar behavior was observed for Apo AI (Bolaños-García et al., 2001).

Experiments on more complex interfaces that are closer to the lipoprotein surface have been prepared adsorbing Apo CI and AII on rac-1,2-dipalmitoyl-sn-glycero-3-phosphocholine (DPPC) monolayers, which indicate that Apolipoproteins can penetrate the DPPC monolayer to form part of the monolayer at the air/water interface (Xicohtencatl-Cortes et al., 2004a). These monolayers also present two clear phase transitions between condensed phases, as well as one between a condensed phase and a gas phase. In this case, the Langmuir monolayer and BAM observations revealed that below surface pressures of 10 mN/m it was possible to have a 2D isotropic mixture where the surface area of the monolayer was approximately the sum of the area occupied respectively by the protein and DPPC molecules as if they were pure components. As the surface pressure is increased and it reaches the condensed phase transition at Π~24-31 mNm⁻¹ there is a important loss of monolayer area with an increasing brightness in one of the condensed phases, as seen with BAM images (see Fig. 5). Taking into account this observations and that the Π values for this condensed transition of the

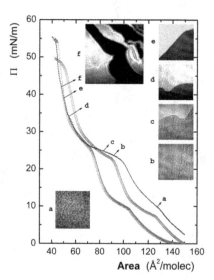
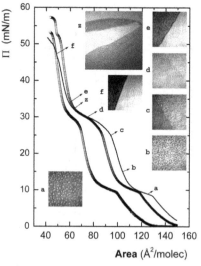

Fig. 5. Langmuir isotherms for Apo CI/DPPC (left, nominal protein mole fraction, from left to right x=0.04, 0.05, 0.12) and Apo AII/DPPC (right, nominal protein mole fraction, from left to right x=0.01, 0.02, 0.03). Insets shows BAM images at different lateral pressures. Adapted from (Xicohtencatl-Cortes et al., 2004a).

binary system are similar to the ones found for the proteins as single components, it was proposed that here there was also a conformational change of the Apo where α-helix segments desorb from the interface, aligning and following the DPPC tails inclination (Xicohtencatl-Cortes et al., 2004a).

4. Forces between adsorbed layers of Apolipoproteins

4.1 The surface force apparatus technique

Forces that control the interaction between proteins, proteins and surfaces and surfaces with adsorbed proteins, are the result of different contributions as hydrophobic interaction, entropy gain due to counterion release, van der Waals force, and to a large extent electrostatic interactions, where the latter is governed by variables like pH and salt concentration. All these interactions depend on the kind of surface and solution where the proteins are immersed, as well as on their charge, shape, and conformation. The surface force apparatus (SFA) (Israelachvili, 1973; Parker et al., 1989) offers the possibility of measuring long-range and contact forces between two mica surfaces covered with adsorbed proteins (Claesson et al., 1995), as well as, measuring the absorbed layer thickness and its compressibility. The latter parameter can give information about the conformational structure and size of the adsorbed protein.

The SFA instrument and experimental procedures have been described by Israelachvili (Israelachvili & McGuiggan, 1990) and Parker (Parker et al., 1989). In general, the force is measured between two curved molecularly smooth mica surfaces (typically 1 cm^2 of area with 2-5 μm constant thickness) where a silver layer of about 520 Å thick was deposited through evaporation on one side of each surface. After that the mica pieces are glued with an epoxy resin, with the silver side down, onto optically polished half-cylindrical silica disks (mean radius of curvature, R, ~ 1-2 cm) that are finally mounted in a crossed cylinder configuration on the SFA. Here, one of the disks is mounted on a double cantilever spring (spring constant, k, ~105 N/m) and the second one on a piezoelectric crystal. This setup produces an optical interferometer. The separation between the two surfaces, d, is controlled by the piezoelectric crystal and the absolute distance is measured interferometrically using fringes of equal chromatic order (FECO) with an accuracy of 2 Å. The magnitude of the force, F, as a function of the surface separation, normalized with respect to the mean radius of curvature, can be determined from the spring deflection measured down to ca. 10^{-7} N. Usually, a SFA experiment starts with the measurement of a standard force curve of water or a buffer solution. If the surfaces contact position is clean and the force curve measured is consistent with the theoretical Derjaguin, Landau, Verwey and Overbeek (DLVO) theory predictions, a known amount of protein is added to the SFA chamber to allow a slow adsorption to the surfaces from the surrounding solution. Then the force curves are usually measured at different times to evaluate this protein adsorption process.

With the SFA, as with other force measurement techniques, one has to consider that the comparison between theoretical and experimental force curves is not straightforward, since the measured force is the sum of different contributions, which are interrelated and therefore not easy to separate. In general, the electrostatic-double layer and the van der Waals forces are considered the most important contributions. However, an absolute determination of the magnitude of each of these forces is complex, due to factors as protein and surface charge density, protein concentration and solution ionic strength, contribution from steric interactions at short distances, etc. In addition, the location of the plane of charge and the dielectric properties of the adsorbed protein layer usually cannot be determined unambiguously. Nevertheless, the results from SFA studies of the interaction between layers of globular proteins, like insulin and lysozyme, and of proteins with disordered structures have increased our knowledge on the proteins adsorbed layer structure (Claesson et al., 1995). This also includes our SFA studies with proteins formed mainly by α-helices, which will be described below.

4.2 Force measurements with Apolipoproteins deposited on hydrophilic surfaces

In general, the force curves measured between hydrophilic surfaces with adsorbed layers of Apos are mainly composed of electrostatic double layer forces at large surface separations and of steric repulsive forces at small distances. These steric forces are quite interesting since they give some insights of the preferred Apos conformations and the interaction produced by them. Apos amphiphatic structure produces a directional adsorption where the hydrophilic faces of the protein α-helices prefer to be adsorbed onto the mica leaving the hydrophobic faces of the α-helices in contact with water. As an example, figure 6 shows the force curves measured, using a SFA, between two mica surfaces adsorbed with Apo AII. In this case, the adsorption was produced from the protein buffer solution (acetic acid-sodium acetate, pH=4) that surrounds the surfaces. Also, a sequential increment of the protein concentration from 0.002 to 0.004 mg/mL was produced to observe the effect in the surface adsorption (Ramos et al., 2008).

As it can be observed, no forces were found until a surface separation of 700 Å was reached. From there, if the surfaces are brought together, a long-range repulsive force is observed until it is overcome by an attractive force (inward jump), which brings the surfaces from a surface separation of about 130-200 Å into a closer contact. The surface separation where the attractive force drives the surfaces close together decreases with adsorption time and it disappears if the protein concentration is increased (see the force curves with 0.004 mg/mL after 8 hrs of adsorption). In some cases, a small repulsive force was found before reaching a repulsive hard wall. The hard wall at the lower concentration was found to be at a surface separation, d, of 11 Å. Interestingly, when the protein concentration was increased (to 0.004 mg/mL) the surface separation value for the hard wall increases with protein adsorption time with approximately 10 Å increments. Also, an adhesive pull-force was found when the surfaces are taken apart, which decreases with protein adsorption. It was also observed that the force curves were the same on compression and on separation if the surfaces are not brought closer than the inward jump. Similar force curves were found for Apo CI (Campos-Terán et al., 2004).

In this case, the long-range repulsive force and the attractive force can be fitted using DLVO theory including additive contributions of non-retarded van der Waals forces and the electrostatic double layer force (see Fig. 6). Calculations of double layer force were performed with the algorithm of Chan et al. (Chan et al., 1980) bringing into play both constant surface potential and constant surface charge. In practice, it is most likely that both potential and surface charge vary as the surfaces approach, where the actual double layer force, as it is in this case, falls between these two limits due to the proteins charge regulation. Although DLVO theory does not take into account additional forces occurring between the surfaces, e. g., hydration forces, hydrophobic forces, and steric forces, etc., the fitting is quite good, and the attractive force measured is close to what theory suggests, at constant surface potential.

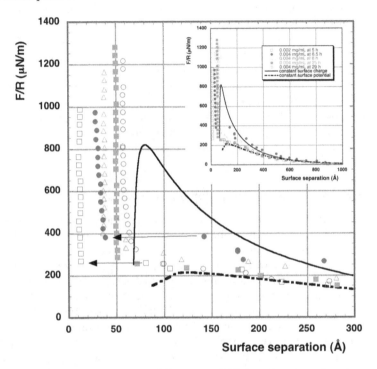

Fig. 6. Force normalized by the radius of curvature, F/R, as a function of surface separation and total adsorption time between mica surfaces adsorbed with Apo AII. The total protein concentration was increased at two times during the experiment and it is: 0.002 mg/mL at 5 h (□), 0.004 mg/ml: at 6 ½ h (●), at 8 h (Δ), and 25 h (■), 29 h (○). Lines indicate DLVO fitting (0.004 mg/ml) with constant surface charge and dashed lines with constant surface potential. Arrows indicate attractive jumps. Adapted from (Ramos et al., 2008).

4.2.1 Short-range forces probe the orientation of the adsorbed protein

As mention before, the analysis of the measured force curves can give an insight of the protein conformation at the surfaces. For the case described above it was seen that the average distance where attractive force appears, d~150 Å, is close to double of the maximum length of this proteins (~85 Å), which suggests that entire protein is oriented perpendicular to the surface or that individual protein segments, i. e, α-helices, protrudes from them. These protein segments could take part in bridging between the two surfaces and thus be responsible for the attractive force. This kind of attractive force was also observed in adsorbed surfaces with Apo CI (Campos-Terán et al., 2004). In addition, studies of Apo AII and Apo CI monolayers have shown that it is possible to form a layer with protruding segments at an interface (Bolaños-García et al., 2001; Ruíz-García et al., 2003). The fact that at short adsorption time, it was observed a weak repulsive force, suggests a more extended conformation of the adsorbed proteins. Such protruding segments could be compressed, bearing in mind the relative flexibility of the polypeptide chains connecting the α-helices. Given enough time for adsorption, the protein molecule preferentially will be oriented parallel to the surface and hence, this repulsive force disappears. The driving force for protein reorientation is to avoid the exposure of hydrophobic segments to the aqueous environment, as well as to promote the electrostatic attractive interactions between the protein and the surface. At low concentrations, the attractive interaction is reduced with time, which implies a higher protein surface coverage confined in a thin layer. This thin protein layer was found experimentally since it was found a surface separation of just 11 Å at the hard wall position. This final layer thickness value is between 5 to 6 Å on each surface, which is similar to the estimated value for α-helices diameter. Previously, it has been shown that structural changes on adsorption are not enough to disturb the α-helix structure (Burkett & Read, 2001).

4.2.2 Sequential addition of Apos builds up protein multilayers

Experiments conducted at higher concentrations suggest the build up of more than one layer on each surface since each curve shown in figure 5 represents an increase in hard-wall separation of ~10 Å or approximately 5 Å of thickness on each surface (see Fig. 6). Confirmation of this process was obtained by ellipsometry measurements done by our group (Ramos et al., 2008), which showed that sequential addition of protein (at least at high ionic strength) leads to an increase in the adsorbed amount of protein, as well as the protein layer thickness. However, for the case of Apo AII, the presence of a repulsive interaction showed that protein adsorption do not lead to charge neutralization of the mica surface charge as it was found for Apo CI (Campos-Terán et al., 2004). This is most likely due to the structural difference between both proteins, where the Apo CI monomer can more efficiently arrange so that it better match the surface charge compare to the Apo AII dimer. However, since the apparent surface potential has a small change when the protein concentration is increased from 0.002 mg/ml to 0.004 mg/ml, a charge regulation mechanism involving small ions during the adsorption of the proteins cannot be discarded. This mechanism has been observed to occur in the surface adsorption of other proteins (Claesson et al., 1995). Protein multilayer adsorption has been observed in proteins with amphiphilic or flexible segments, as observed at SFA experiments with cytochrome c (Kekicheff et al., 1990) and β-casein (Nylander & Wahlgren, 1997). A multilayer protein adsorption requires attractive protein-protein interaction, which often is weaker than the protein-surface interactions. Confirmation of this statement was obtained by diluting the solution surrounding the surfaces. Here, the hard wall separation decreased from d ≈ 58 Å to d ≈ 26 Å and the apparent surface potential has increased from ~37 mV to ~53 mV 1 ½ h after dilution, which indicates protein desorption. Even more protein has desorbed after 18 h, and the hard-wall separation reaches d ≈ 11 Å, corresponding to one monolayer on each surface. In addition, an attractive jump appears. No further desorption occurs, which is mostly likely due to the strong interaction between the negatively charged mica and the cationic protein as well as the entropy gain due to counter ion release. This experiment showed the reversibility of Apo AII adsorption process that produces in each stage different protein conformations. Also, it is noteworthy that quite similar force curves were observed in Apo CI (Campos-Terán et al., 2004), which also has a similar secondary structure but with a different peptide sequence, different net charge, and it is only monomeric. Therefore, the observed force curves seem to be a consequence of the particular features of the amphiphilic α-helices.

5. Lipid dependant disorder-to-order conformational transitions in Apolipoproteins

5.1 Apo CI derived peptides-lipid interaction

As mentioned in sections 3 and 4, it has been observed for exchangeable Apolipoproteins that when they are subjected to lateral pressures then several helical segments were placed directly in the hydrophobic phase of the interface. In the case of Apo CI, our data showed an interesting new property since we observed that injecting it into the subphase allows the protein to go to the water/lipid interface quickly and when lateral pressure is increased the C-terminal helical segment penetrates the monolayer. In addition, when lateral pressure is released, this segment is again incorporated into the water/lipid interface (Bolaños-García et al., 1999). With these results we were interested to know if the secondary structure of the C-terminal segment of Apo CI remained stable regardless of their position in the different hydrophilic/hydrophobic interfaces. To solve this question we conducted studies of peptides derived from the C-terminal segment of Apo CI in different environments.

The peptides were designated according to the first three letters of their amino acid sequence and called ALDO (A7-E24), ARELI (A22-M38) and SAK (S35-L53). Apo CI in solution shows a clear circular dichroism (CD) signal associated with a high degree of α-helix structure (Bolaños-García et al., 1999). However, when peptides ALDO, ARELI and SAK (Mendoza-Espinosa et al., 2008) were tested under the same experimental conditions, they showed no defined secondary structure and remain non-structured independently of pH, temperature and ionic strength. Interestingly, despite that these peptides have an amphipathic character and high hydrophobic moment values ($\mu H > 0.315$ kcal/mol), they remain completely unfolded in solution (see Fig. 7a).

Nevertheless, when peptides ALDO, SAK and ARELI are placed in aqueous solution with 40% v/v trifluoroethanol (TFE) or sodium dodecylsulphate (SDS, cmc of 8.5 mM) they show a CD signal clearly associated with an α-helical structure. If SDS was used at different concentrations (1.5-20 mM), each of the peptides acquire secondary structures in a differentiated way, where the lowest percentage of α-helix structure corresponds to ARELI and the highest to SAK peptide, which corresponds to the C-terminal segment of Apo CI (see Figs. 7b and 7c). Then in order to test the possibility that specific lipids on the surface of lipoproteins and plasma membrane induce an α-helix conformation as in the case of TFE and SDS, we tested a series of phospholipids above and below its critical micelle concentration and with different acyl long chain to probe their hydrophobic effect. L-α-Phosphatidylcholine (PC) was used above its critical micellar concentration (cmc <0.005 mM) and 1,2-dihexanoyl-sn-glycero-3-phosphocholine (DHPC) slightly below its cmc (~15 mM), because concentrations above the cmc of DHPC generate solutions that prevent the determination of the CD signal (data not shown). Under these conditions, only peptide SAK showed a well-defined disorder to order type transition. Since medium hydrophobicity seems to be critical for the transition to be observed, we tested if these lipids mixed with small amounts of cholesterol altered these low percentages of α-helix, finding no changes (data not shown).

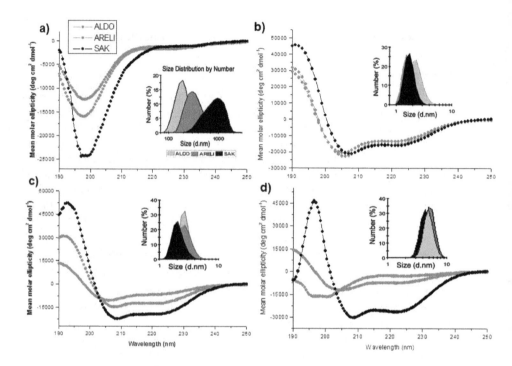

Fig. 7. Far-UV CD data of Apo CI-derived peptides. a) Spectra recorded in water for peptides ALDO, ARELI, and SAK. b) In the presence of 40% v/v TFE. c) In the presence of SDS (20 mM). d) In the presence of lyso-C_{12}PC (20 mM). Insets: DLS analysis of the same corresponding peptide solutions employed for CD experiments. Adapted from (Mendoza-Espinosa et al., 2008).

Tests performed with 1-hexanoyl-2-hydroxy-sn-glycero-3-phosphocholine (lyso-C_6PC) do not promote any change in the secondary structure of all peptides studied (data not shown). However, in the presence of 1-lauroyl-2-hydroxy-sn-glycero-3-phosphocholine (lyso-C_{12}PC), peptides corresponding to the N-terminal (ALDO) and C-terminal (SAK) segments acquired different percentages of α-helix structure as a function of the lysophospholipid concentration. ALDO was the peptide less sensitive to lyso-C_{12}PC, however, the effect of this lysophospholipid is greater than that generated by SDS. On the other hand, SAK presented disorder-to-order transitions from the lowest levels of lyso-C_{12}PC. Greater effect was observed for this lipid on this peptide compared to the one observed with SDS and TFE molecules (see Fig. 7d). Interestingly DLS experiments showed that peptide solutions with pure water and lyso-C_6PC in which there was no disorder-order transitions, presented aggregates in solution. In contrast, for the peptides-lyso-C_{12}PC solutions that generated disorder-to-order transition and the promotion of a well-defined α-helical conformation, allows the association of lipid/peptide molecules in such an orderly fashion that the system avoids aggregation. It is interesting to note that while lyso-C_6PC aggregates increase in size in the presence of peptides for the case of lyso-C_{12}PC the size does not change with or without the same peptides (see Fig. 8).

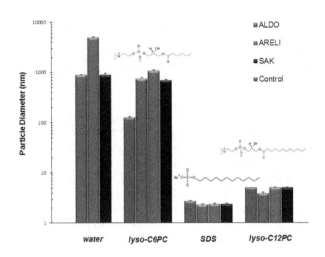

Fig. 8. Dynamic light scattering of Apo CI peptides associated with lysophospholipids of different acyl-chain lengths. a) Quantification of particle diameters given by peptide/lipid aggregates in the presence of SDS (20mM), lyso-C_6PC (20 mM) and lyso-C_{12}PC (20 mM). Adapted from (Mendoza-Espinosa et al., 2008).

Based on data obtained in this study, we have elucidated a mechanism by which the Apo CI could be functioning as a molecular switch on the surface of HDL. In this scenario we propose that Apo CI responds to a decrease in lateral pressure on the surface of HDL, which is given by an increase of cholesterol ester in the nucleus (Frank & Marcel, 2000), by promoting its C-terminal segment to the polar/nonpolar interface of the lipoprotein particle with a concomitant change from a disordered structure to an α-helix. The fact that the surface of lipoproteins and certain types of membranes are associated with the presence of molecules such as lyso-C_{12}PC could generate dramatic disorder-to-order transitions in the C-terminal segment of Apo CI. In consequence, these conformational changes generated by Apo CI could be related to the biological activity of molecules such as esphingosine 1 phosphate that when associated with HDL particles it has been observed that promotes an anti-inflammatory effect and therefore presents a potential role as atheroprotective (Jerzy-Roch & Assman, 2005).

Also, since cholesterol esters formed by the enzyme lecithin-cholesterol acyltransferase (LCAT) located at the surface of HDL particles promotes the transfer of a fatty acyl group from position two of phosphatidylcholine to cholesterol, with the consequent synthesis of lysophosphatidylcholine, it is possible that the presence of new OH groups at the polar/non-polar interface change the electrostatic properties of the interface and the way water is displaced from the interface during peptide folding. In fact, it has been proposed that in the presence of lipids, the process of peptide folding corresponds to an enthalpy driven process supported by the energy employed for water displacement (Rozek et al., 1997). Localized changes in secondary structure of a number of proteins have been found to be physiologically relevant (Chakrabartty & Baldwin, 1995; Meador et al., 1992). Therefore, a series of conformational switches have been proposed, in specific cases, to promote protein activation (Wei et al., 1994) and folding (Hamada et al., 1996). In order to find out the mechanism by which lyso-C_{12}PC is required to induce an important conformational change, further investigation is needed. These important changes might be important in the understanding of the mechanisms Apo CI employs to modulate protein/protein recognition directly related to enzyme activation and modulation of Apo E and the cholesterol ester transfer protein (CETP) function when associated to the surface of HDL particles. Our proposal of a lipid dependant disorder-to-order conformational transition in Apo CI might be considered a conformational switch mediating enzyme activation and lipid transport. This possibility opens new ways to visualize the concert of events that take place at the surface of HDL during their transformation from early protein/lipid discoidal aggregates to spherical particles, ready to be taken up by liver cells. Further investigation of this potential mechanism designed to recognize and promote localized secondary structure conformations in proteins, undoubtedly will provide an improvement to better comprehend the protein function at the surface of lipoproteins.

5.2 Apo AI-lipid interactions

Several studies have evaluated the lipid-binding propensity of each of the helices composing Apo AI, noting that the N-terminal domain determines the open or closed structure of the protein when modulated by the presence of cholesterol obtained by interacting with the ATP-binding cassette (ABC) A1 (ABCA1) giving rise to the nascent HDL. Likewise, due to its high hydrophobicity, the C-terminal domain of Apo AI facilitates anchoring to lipid membranes (Fang et al., 2003; Kono et al., 2008). These studies are based on the widely accepted model for discoidal HDL, which corresponds to a disk made of a lipid bilayer surrounded by two Apo AI helices with its long axis perpendicular to the acyl chains of phospholipids (Garda, 2007). These properties can be easily observed in the hydrophobicity profile of Apo AI obtained with the use of the EMBOSS Pepinfo algorithm, employing a window of 9 amino acids and the scale of Kyte J. & Doolittle R. F. (Kyte & Doolittle, 1982) (see Fig. 9a). While its negative profile is characteristic of membrane proteins at the N and C-terminal regions of the protein (10-17, 213-229), the positive profile indicates the hydrophobic ones. On the other hand, the use of the Hydrophobic Cluster Analysis (HCA) server, which predicts hydrophobic blocks depending on the secondary structure of the polypeptide chain, shows three highly hydrophobic segments (aa 13-22, 45-49, 216-232) (Fig. 9a, hydrophobic clusters) (Callebaut et al., 1997). The distribution in the helix of negatively charged, positive or neutral aminoacids, generates the different types of helices present in the Apo AI. This change in the distribution of amino acids is important in the understanding of the way the protein associates with lipids (Segrest et al., 1992). For example, segments corresponding to helices 1-2 (A-type helices) and 9-10 (Y-type helices) are those with the greatest affinity for lipids, which are particularly high in the latter (Mishra et al., 1998).

Interestingly, in two SDSL-EPR spectroscopic studies (electron paramagnetic spin-label resonance spectroscopy), β-type segments were also detected in the N-and C-terminal domains of Apo AI (Lagerstedt et al., 2007; Oda et al., 2003). The possibility of having secondary structure conformational changes has been also observed in other proteins. For instance, the fusogenic HA2 unit of hemagglutinin of the influenza virus has been shown to present these types of conformational transitions. HA2 corresponds to a segment containing 36 amino acids, that presents the ability to carry out transitions from a random coil structure to an α-helix domain. The presence of these secondary structure conformational changes in Apo AI in the presence of lipid could serve as a mechanism to decrease the energy barrier in their interaction with these molecules, a crucial step in the flow of cholesterol and assembly of HDL (Oda et al., 2003; Tamm, 2003).

5.3 Intrinsic disorder in Apo AI

On the other hand, Apo AI is considered within the group of natively unstructured proteins (Uversky et al., 2000). Recently, this type of protein has taken a major importance when giving rise to the term "unfoldomics". A highly dysfunctional group of proteins has been associated with a number of conditions such as amyloidosis, cancer, diabetes, neurodegenerative diseases and others. The altered sites contained in many disordered proteins have been shown to be highly susceptible to proteolysis. In the lipid-free Apo AI, specifically the N-terminal segment has been observed by various techniques such as NMR, EPR that mobility presents a great variability in their secondary structure (Kono et al., 2008; Lagerstedt et al., 2007; Okon et al., 2001, 2002; Wang et al., 1996, 1997). Using the PONDR server with the native sequence of the native Apo AI it was estimated a high percentage of disorder for five segments of Apo AI (Fig. 9b). The first segment (aa 1-10) corresponds to a site that could serve as a lipid sensor when the Apo AI is in the discoidal particle (Kono et al., 2008; Wu et al., 2007). The second site (aa 69-89) has a particularly negative charge distribution compared to the other un-structured sites (Fig 9c). The site also includes a transition between a helix type A to a type Y, which has been postulated could be a destabilizing factor in the continuity of the secondary structure of an α-helix. Wu Z. et al (Wu et al., 2007) have proposed a discoidal HDL model where the third and longest disordered segment (aa 116-150) presents a region that could be considered as a hinge. The same model includes a loop (159-180) that corresponds to the fourth disordered segment (aa 172-194). Also, this site is adjacent to the segment postulated to be the one that interacts with LCAT (aa 159-170). The fifth disordered segment is located close to a transition region from a helix type A to a type Y. Although the latter site presents an α-helix structure in the crystal structure of Apo AI, by EPR tests this segment shows a β-structure that could serve as a mechanism to facilitate the interaction with lipids. Interestingly, very low concentrations of amyloid fibrils formed by a segment of 10 kD N-terminus of native Apo AI, have been found in vivo (Schmidt et al., 1997). These structures are constituted by β-cross structures that in turn produce β-strands oriented in a perpendicular way with respect to the long axis of the fiber, resulting in its increased spreading capacity. Subsequently, protofibrils associate laterally or rotate together to form fibers of larger diameter (DuBay et al., 2004).

In our laboratory by their structural and perhaps biological importance, we have analyzed three peptides designed according to the sequence reported for the native Apo AI and its crystal structure (Borhani et al., 1997). These peptides are DRV (D9-D24) and KLL (K45-Q63) located within its N-terminal helical segments, and VLES (V221-K239) located in a C-terminal segment. By sequence analysis of native Apo AI employing the Zyggregator server (Tartaglia et al., 2008), it has been observed that this protein presents several sites with the propensity to form amyloid fibrils (A15-D20, W50-F57 and S224-L230), which interestingly enough are included in peptide sequence DRV, KLL and VLES (Fig. 9d, Zagg Propensity). This server uses an algorithm that considers patterns of hydrophobicity, polar amino acids

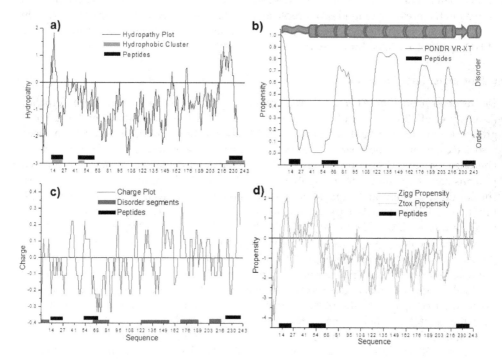

Fig. 9. Methods employed in the prediction of disorder, aggregation and propensity of amyloid fiber formation based on the sequence of native Apo AI. a) Hydrophobicity profile calculated with the EMBOSS server: Pepinfo, using a window of 9 aa and scale of Kyte & Doolittle hydrophobicity, red line. Hydrophobic segments calculated with the HCA server, orange box. Peptide position DRV, KLL and VLES in the sequence of Apo AI, black box. b) Profile of disorder determined by the PONDR server (PONDR VL-XT). c) Load profile calculated with the EMBOSS server: Charge using a window of 9 aa. d) Propensity of amyloid fiber formation with the Zyggregator server (Zagg Propensity) and formation of globular structures (Ztox Propensity).

and aromatic content observed in polypeptide chains of amyloidogenic proteins. The segments prone to aggregation in the Apo AI present sequences of highly hydrophobic blocks composed of six to seven amino acids flanked by negative and/or positive charges. On the other hand, Zyggregator calculates the tendency to form globular structures, which have been observed to be a step in the formation of amyloidogenic fibers. It has been observed with in vitro experiments that these globular structures formed by amyloid peptides are involved in a process of cellular cytotoxicity due the formation of pores in membranes (Lashuel & Lansbury, 2006). The patterns to form amyloid structures in Apo AI calculated by Zyggregator, show the same one that is observed during the formation of amyloid fibers (Fig. 9d, Ztox propensity).

5.4 Amyloidogenic Apo AI

Interestingly, within the existing mutants of Apo AI, there are 4 isoforms associated with systemic forms of hereditary amyloidosis. Mutations correspond to Gly26-Arg, Leu60-Arg, Trp50-Arg and a deletion/insertion of segment (Leu60-Phe71) - (Val-Val-Thr). Considering the structure for the lipid-free Apo AI proposed by Ajees et al., these mutations generally provide positive charges to the hydrophobic interface formed between the two pairs of helices at the N-terminal segment (aa 1-188). The introduction of a polar amino acid residue by the amyloidogenic mutations in the hydrophobic interface of the lipid-free Apo AI, probably prevents the formation of the cluster of α-helices in the N-terminal structure. This also hinders the formation of hydrophilic patches located in different areas of the protein (see Fig. 10) (Oram, 2002). These hydrophilic patches have been postulated to interact with ABCA1 for the transfer of phospholipid and cholesterol. One consequence of this obstacle might be that the formation of a properly sized discoidal HDL needed to interact with the enzyme LCAT, as observed in these mutations, is not properly achieved (Fang et al., 2003; Genschel et al., 1998). This interaction is crucial in the transition from discoidal to spherical HDL (Calabresi & Franceschini, 2010). Discoidal HDL formed by one of the several isoforms of the amyloidogenic Apo AI known nowadays, are rapidly catabolized and do not become spherical HDL (Genschel et al., 1998). At this stage, it is interesting to mention that the metabolic pathway of these Apo AI isoforms that cause deposition of amyloid fibrils, has not yet been clarified. However any of these mutations found in helix 1 and helix G * of Apo AI could be generating a loop susceptible to proteolysis between helices 2 and 3 (Apo AI disordered second site with low affinity to lipid profile and with a distinctly negative charge) (Figs. 9b and 9c). In all cases, amyloid fibers are generated with polypeptides from the first 83-94 amino acids at the N-terminus of Apo AI. Mutations at amino acids 26, 50 and 60 also generate charge changes, characteristic that favors the formation of extended β-sheets (García-González & Mas-Oliva, 2011). These peptides released regardless of the origin of the mutation, always have the same net charge, indicating the conservation of the hydrophilic profile. Likewise, the hydrophobic moment value and the average total hydrophobicity decrease in the mutations with respect to native sequences.

It is remarkable that several theoretical and experimental data related to several N-terminal and C-terminal segments of Apo AI indicate that both have a high propensity for aggregation. However, only the first one was found in the amyloidogenic plaques isolated from familial amyloidosis or Alzheimer-affected people. Segments of Apo AI with a tendency to be maintained in a disordered state, together with their low affinity sites for lipid, could be the key for the understanding of HDL particles formation. Due to these characteristics, it seems Apo AI has the ability to modulate its secondary structure based on the presence of hydrophobic/hydrophilic interfaces that in turn might activate or inhibit the function of proteins that regulate the metabolism of HDL.

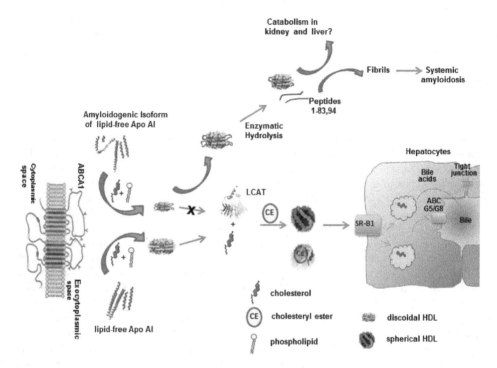

Fig. 10. The ABCA1 receptor promotes the transfer of phospholipids to a lipid-poor form of native Apo AI native, the major component of HDLs. The mechanism by which this process occurs is not fully understood; probably ABCA1 translocates phospholipids and cholesterol from the plasma membrane to HDL Apo AI producing a discoidal particle. These discoidal particles are transformed into spherical HDLs in the blood by the action of the enzyme LCAT. Spherical HDLs associate with the SRB1 receptor in the plasma membrane of hepatocytes and transfer free and esterified cholesterol to the liver for excretion into the bile as free cholesterol (via ABCG5/G8) or subsequent conversion to bile salts (Oram, 2002). In the case of the formation of amyloidogenic isoforms of Apo AI, a proper interaction with ABCA1 is impeded and therefore the formation of discoidal HDLs becomes a deficient process, with the consequent problems in the recognition of LCAT and also SR-B1. It is known that abnormal HDLs are rapidly catabolized releasing different peptides that in turn generate peptides that form amyloid fibers that might initiate a systemic amyloidosis.

6. Conclusions

We believe conformational changes observed in monolayers of Apo CI and AII during lateral compression could be of direct relevance to changes in surface tension at the surface of lipoproteins. Thus, in response to changes in surface tension of the lipoprotein particle, Apolipoproteins could present structural changes. Interestingly, our working hypothesis is of relevance when extrapolated to the conformational changes observed at the lipid/plasma interface at the surface of a lipoprotein particle. In this case, because of the small size and rich protein composition of discoidal HDL (pre-2HDL β) it is proposed that the lateral pressure in the phospholipid monolayer of these particles is most likely to be high and only decreases in parallel with changes in size and shape of these lipoproteins when they begin to accumulate cholesterol esters to form HDL spherical (α-HDL). The results of our studies employing Apos allow us to postulate that the lateral pressure of the phospholipids monolayer associated with proteins on the surface of the different HDL particles may be very different depending on their size and shape.

In addition, the fact that these Apos could present a different conformation from newly synthesized lipoproteins with a discoidal form to a mature state with a spherical shape could be an important factor in the understanding of their physiological properties such as directionality and receptor recognition. However, the ways in which structural changes induced by lipid interaction modulate the functionality of these Apos are still to be clarified, since the formation of amyloidogenic forms for several segments of these Apos, as presented in this chapter, have also been found to play a critical role in their structure/function relationship.

Currently, the understanding of the mechanisms by which segments, entire native or mutated proteins get transformed into amyloid-like structures, has resulted to be a challenge. Since several disorder-to-order transitions in proteins have been found to be reversible, this phenomenon has been frequently associated with important signaling events in the cell. Due to the central role of this phenomenon in cell biology, protein misfolding and aberrant conformational transitions have been at present associated with an important number of diseases. Nevertheless, differences between "functional" and "pathological" disorder-to-order or order-to-disorder transitions that might lead to the formation of amyloids, might simply reside in the modulatory pathways involved along their synthesis and the environment proteins or protein segments are placed into.

7. Acknowledgment

Research described in the present chapter has been supported by Consejo Nacional de Ciencia y Tecnología (CONACYT), Red Temática de Materia Condensada Blanda (CONACYT), DGAPA-UNAM (Universidad Nacional Autónoma de México) and UAM (Universidad Autónoma Metropolitana).

8. References

Ajees, A.A., Anantharamaiah, G.M., Mishra, V.K., Hussain, M.M. & Murthy, H.M. (2006). Crystal Structure of Human Apolipoprotein A-I: Insights into Its Protective Effect against Cardiovascular Diseases. *Proceedings of the National Academy of Sciences of the United States of America*, Vol. 103, No. 7, (February 2006), pp. 2126-2131, ISSN 0027-8424

Andreola, A., Bellotti, V., Giorgetti, S., Mangione, P., Obici, L., Stoppini, M., Torres, J., Monzani, E., Merlini, G. & Sunde, M. (2003). Conformational Switching and

Fibrillogenesis in the Amyloidogenic Fragment of Apolipoprotein A-I. *The Journal of Biological Chemistry*, Vol. 278, No. 4, (January 2003), pp. 2444-2451, ISSN 0021-9258

Bode, W. & Huber, R. (1976). Induction of the Bovine Trypsinogen-Trypsin Transition by Peptides Sequentially Similar to the N-Terminus of Trypsin. *Federation of European Biochemical Societies Letters*, Vol. 68, No. 2, (October 1976), pp. 231-236, ISSN 0014-5793

Bolaños-García, V.M., Mas-Oliva, J., Ramos, S. & Castillo, R. (1999). Phase Transitions in Monolayers of Human Apolipoprotein C-I. *Journal of Physical Chemistry B*, Vol. 103, No. 30, (July 1999), pp. 6236-6242, ISSN 1089-5647

Bolaños-García, V.M., Ramos, S., Xicohtencatl-Cortes, J., Castillo, R. & Mas-Oliva, J. (2001). Monolayers of Apolipoproteins at the Air/Water Interface. *Journal of Physical Chemistry B*, Vol. 105, No. 24, (June 2001), pp. 5757-5765, ISSN 1089-5647

Bolaños-Garcia, V.M., Soriano-Garcia, M. & Mas-Oliva, J. (1997). CETP and Exchangeable Apoproteins: Common Features in Lipid Binding Activity. *Molecular and Cellular Biochemistry*, Vol. 175, No. 1-2, (October 1997), pp. 1-10, ISSN 0300-8177

Borhani, D.W., Rogers, D.P., Engler, J.A. & Brouillette, C.G. (1997). Crystal Structure of Truncated Human Apolipoprotein A-I Suggests a Lipid-Bound Conformation. *Proceedings of the National Academy of Sciences of the United States of America*, Vol. 94, No. 23, (November 1997), pp. 12291-12296, ISSN 0027-8424

Brewer, H.B., Jr., Lux, S.E., Ronan, R. & John, K.M. (1972). Amino Acid Sequence of Human ApoLp-Gln-II (ApoA-II), an Apolipoprotein Isolated from the High-Density Lipoprotein Complex. *Proceedings of the National Academy of Sciences of the United States of America*, Vol. 69, No. 5, (May 1972), pp. 1304-1308, ISSN 0027-8424

Brewer, H.B., Jr., Ronan, R., Meng, M. & Bishop, C. (1986). Isolation and Characterization of Apolipoproteins A-I, A-II, and A-IV. *Methods In Enzymology*, Vol. 128, (1986), pp. 223-246, ISSN 0076-6879

Burkett, S.L. & Read, M.J. (2001). Adsorption-Induced Conformational Changes of α-Helical Peptides. *Langmuir: The ACS Journal of Surfaces and Colloids*, Vol. 17, No. 16, (July 2001), pp. 5059-5065, ISSN 0743-7463

Bustos, D.M. & Iglesias, A.A. (2006). Intrinsic Disorder is a Key Characteristic in Partners that Bind 14-3-3 Proteins. *Proteins*, Vol. 63, No. 1, (April 2006), pp. 35-42, ISSN 1097-0134

Calabresi, L. & Franceschini, G. (2010). Lecithin:Cholesterol Acyltransferase, High-Density Lipoproteins, and Atheroprotection in Humans. *Trends in Cardiovascular Medicine*, Vol. 20, No. 2, (February 2010), pp. 50-53, ISSN 1873-2615

Callebaut, I., Labesse, G., Durand, P., Poupon, A., Canard, L., Chomilier, J., Henrissat, B. & Mornon, J.P. (1997). Deciphering Protein Sequence Information through Hydrophobic Cluster Analysis (HCA): Current Status and Perspectives. *Cellular and Molecular Life Sciences : CMLS*, Vol. 53, No. 8, (August 1997), pp. 621-645, ISSN 1420-682X

Campos-Terán, J., Mas-Oliva, J. & Castillo, R. (2004). Interactions and Conformations of α-Helical Human Apolipoprotein CI on Hydrophilic and on Hydrophobic Substrates. *Journal of Physical Chemistry B*, Vol. 108, No. 52, (November 2004), pp. 20442-20450, ISSN 1089-5647

Castro, G.R. & Fielding, C.J. (1984). Evidence for the Distribution of Apolipoprotein E between Lipoprotein Classes in Human Normocholesterolemic Plasma and for the Origin of Unassociated Apolipoprotein E (Lp-E). *Journal of Lipid Research*, Vol. 25, No. 1, (January 1984), pp. 58-67, ISSN 0022-2275

Chakrabartty, A. & Baldwin R.L (1995). Stability of α-Helices. *Advances in Protein Chemistry*, Vol. 46, (1995), pp. 141-176, ISSN 0065-3233

Chan, D.Y.C., Pashley, R.M. & White, L.R. (1980). A Simple Algorithm for the Calculation of the Electrostatic Repulsion between Identical Charged Surfaces in Electrolyte. *Journal of Colloid and Interface Science*, Vol. 77, No. 1, (September 1980), pp. 283-285, ISSN 0021-9797

Claesson, P.M., Blomberg, E., Fröberg, J.C., Nylander, T. & Arnebrant, T. (1995). Protein Interactions at Solid Surfaces. *Advances in Colloid and Interface Science*, Vol. 57, (May 1995), pp. 161-227, ISSN 0001-8686

Clay, M.A., Cehic, D.A., Pyle, D.H., Rye, K.A. & Barter, P.J. (1999). Formation of Apolipoprotein-Specific High-Density Lipoprotein Particles from Lipid-Free Apolipoproteins A-I and A-II. *Biochemical Journal*, Vol. 337 (February 1999), pp. 445-451, ISSN 0264-6021

Conchillo-Solé, O., de Groot, N.S., Aviles, F.X., Vendrell, J., Daura, X. & Ventura, S. (2007). Aggrescan: A Server for the Prediction and Evaluation of "Hot Spots" of Aggregation in Polypeptides. *BMC Bioinformatics*, Vol. 8, (February 2007), pp. 65, ISSN 1471-2105

Cortese, M.S., Uversky, V.N. & Dunker, A.K. (2008). Intrinsic Disorder in Scaffold Proteins: Getting More from Less. *Progress in Biophysics and Molecular Biology*, Vol. 98, No. 1, (September 2008), pp. 85-106, ISSN 0079-6107

Dalal, S. & Regan, L. (2000). Understanding the Sequence Determinants of Conformational Switching Using Protein Design. *Protein Science : A Publication of the Protein Society*, Vol. 9, No. 9, (September 2000), pp. 1651-1659, ISSN 0961-8368

Despres, J.P., Lemieux, I., Dagenais, G.R., Cantin, B. & Lamarche, B. (2000). HDL-Cholesterol as a Marker of Coronary Heart Disease Risk: The Quebec Cardiovascular Study. *Atherosclerosis*, Vol. 153, No. 2, (December 2000), pp. 263-272, ISSN 0021-9150

Dobson, C.M. (1999). Protein Misfolding, Evolution and Disease. *Trends in Biochemical Sciences*, Vol. 24, No. 9, (September 1999), pp. 329-332, ISSN 0968-0004

Dobson, C.M. (2003). Protein Folding and Misfolding. *Nature*, Vol. 426, No. 6968, (December 2003), pp. 884-890, ISSN 1476-4687

Dobson, C.M. (2004). Protein Chemistry. In the Footsteps of Alchemists. *Science*, Vol. 304, No. 5675, (May 2004), pp. 1259-1262, ISSN 1095-9203

Dubay, K.F., Pawar, A.P., Chiti, F., Zurdo, J., Dobson, C.M. & Vendruscolo, M. (2004). Prediction of the Absolute Aggregation Rates of Amyloidogenic Polypeptide Chains. *Journal of Molecular Biology*, Vol. 341, No. 5, (August 2004), pp. 1317-1326, ISSN 0022-2836

Dunker, A.K., Garner, E., Guilliot, S., Romero, P., Albrecht, K., Hart, J., Obradovic, Z., Kissinger, C. & Villafranca, J.E. (1998). Protein Disorder and the Evolution of Molecular Recognition: Theory, Predictions and Observations. *Pacific Symposium on Biocomputing*, Vol. 3, (August 1998), pp. 473-484, ISSN 1793-5091

Dunker, A.K., Lawson, J.D., Brown, C.J., Williams, R.M., Romero, P., Oh, J.S., Oldfield, C.J., Campen, A.M., Ratliff, C.M., Hipps, K.W., Ausio, J., Nissen, M.S., Reeves, R., Kang, C., Kissinger, C.R., Bailey, R.W., Griswold, M.D., Chiu, W., Garner, E.C. & Obradovic, Z.J. (2001). Intrinsically Disordered Protein. *Journal of Molecular Graphics and Modelling*, Vol. 19, No. 1, (February 2001), pp. 26-59, ISSN: 1093-3263

Eaton, W.A., Muñoz, V.,Hagen, S.J., Jas, G.S., Lapidus, L.J. & Henry, E.R. (2000). Fast Kinetics and Mechanisms in Protein Folding. *Annual Review of Biophysics and Biomolecular Structure*, Vol. 29, (June 2000), pp. 327–359, ISSN: 1056-8700

Eggerman, T.L., Hoeg, J.M., Meng, M.S., Tombragel, A., Bojanovski, D. & Brewer, H.B., Jr. (1991). Differential Tissue-Specific Expression of Human ApoA-I and ApoA-II. *Journal of Lipid Research*, Vol. 32, No. 5, (May 1991), pp. 821-828, ISSN 0022-2275

Eisenberg, D., Nelson, R., Sawaya, M.R., Balbirnie, M., Sambashivan, S., Ivanova, M.I., Madsen, A.O. & Riekel, C. (2006). The Structural Biology of Protein Aggregation Diseases: Fundamental Questions and Some Answers. *Accounts of Chemical Research*, Vol. 39, No. 9, (September 2006), pp. 568-575, ISSN 0001-4842

Eisenberg, S. (1990). Metabolism of Apolipoproteins and Lipoproteins. *Current Opinion in Lipidology*, Vol. 1, No. 3, (June 1990), pp. 205-215, ISSN 0957-9672

Fang, Y., Gursky, O. & Atkinson, D. (2003). Lipid-Binding Studies of Human Apolipoprotein A-I and its Terminally Truncated Mutants. *Biochemistry*, Vol. 42, No. 45, (November 2003), pp. 13260-13268, ISSN 0006-2960

Fink, A.L. (1998). Protein Aggregation: Folding Aggregates, Inclusion Bodies and Amyloid. *Folding & Design*, Vol. 3, No. 1, (February 1998), pp. R9-23, ISSN 1359-0278

Frank, P.G. & Marcel, Y.L. (2000). Apolipoprotein A-I: Structure-Function Relationships. *Journal of Lipid Research*, Vol. 41, No. 6, (June 2000), pp. 853-872, ISSN 0022-2275

García-González, V. & Mas-Oliva, J. (2011). Amyloidogenic Properties of a D/N Mutated 12 Amino Acid Fragment of the C-Terminal Domain of the Cholesteryl-Ester Transfer Protein (CETP). *International Journal of Molecular Sciences*, Vol. 12, No. 3, (March 2011), pp. 2019-2035, ISSN 1422-0067

Garda, H.A. (2007). Structure–Function Relationships in Human Apolipoprotein A-I: Role of a Central Helix Pair. *Future Lipidology*, Vol. 2, No. 1, (February 2007), pp. 95-104, ISSN 1746-0875

Genschel, J., Haas, R., Pröpsting, M.J. & Schmidt, H.H.J. (1998). Hypothesis. Apolipoprotein A-I Induced Amyloidosis. *Federation of European Biochemical Societies Letters*, Vol. 430, No. 3, (July 1998), pp. 145-149, ISSN 0014-5793

Gsponer, J. & Vendruscolo, M. (2006). Theoretical Approaches to Protein Aggregation. *Protein and Peptide Letters*, Vol. 13, No. 3, (March 2006), pp. 287-293, ISSN 0929-8665

Hamada, D., Segawa, S. & Goto, Y. (1996). Non-Native Alpha-Helical Intermediate in the Refolding of Beta-Lactoglobulin, a Predominantly Beta-Sheet Protein. *Nature Structural Biology*, Vol. 3, No. 10, (October 1996), pp. 868-873, ISSN 1072-8368

Huang, K. (2005). *Lectures on Statistical Physics and Protein Folding*, World Scientific Publishing Company, ISBN 978-981-256-143-5, Singapore.

Huber, R. & Bode, W. (1978). Structural Basis of the Activation and Action of Trypsin. *Accounts of Chemical Research*, Vol. 11, No. 3, (March 1978), pp. 114-122, ISSN 0001-4842

Israelachvili, J.N. (1973). Thin Film Studies Using Multiple-Beam Interferometry. *Journal of Colloid and Interface Science*, Vol. 44, No. 2, (August 1973), pp. 259-272, ISSN 0021-9797

Israelachvili, J.N. & McGuiggan, P.M. (1990). Adhesion and Short Range Force between Surfaces. Part I: New Apparatus for Surface Force Measurements. *Journal of Materials Research*, Vol. 5, No. 10, (October 1990), pp. 2223-2231, ISSN 0884-2914

James, L.C. & Tawfik, D.S. (2003). Conformational Diversity and Protein Evolution--a 60-Year-Old Hypothesis Revisited. *Trends in Biochemical Sciences*, Vol. 28, No. 7, (July 2003), pp. 361-368, ISSN 0968-0004

Jerzy-Roch, N. & Assmann, G. (2005). Atheroprotective Effects of High-Density Lipoprotein-Associated Lysosphingolipids. *Trends in Cardiovascular Medicine*, Vol. 15, No. 7, (October 2005), pp. 265–271, ISSN 1050-1738

Kekicheff, P., Ducker, W.A., Ninham, B.W. & Pileni, M.P. (1990). Multilayer Adsorption of Cytochrome c on Mica around Isolectric pH. *Langmuir*, Vol. 6, No. 11 (November 1990), pp. 1704-1708, ISSN 0743-7463

Kono, M., Okumura, Y., Tanaka, M., Nguyen, D., Dhanasekaran, P., Lund-Katz, S., Phillips, M.C. & Saito, H. (2008). Conformational Flexibility of the N-Terminal Domain of Apolipoprotein A-I Bound to Spherical Lipid Particles. *Biochemistry*, Vol. 47, No. 43, (October 2008), pp. 11340-11347, ISSN 1520-4995

Kriwacki, R.W., Hengst, L., Tennant, L., Reed, S.I. & Wright, P.E. (1996). Structural Studies of P21waf1/Cip1/Sdi1 in the Free and Cdk2-Bound State: Conformational Disorder Mediates Binding Diversity. *Proceedings of the National Academy of Sciences of the United States of America*, Vol. 93, No. 21, (October 1996), pp. 11504-11509, ISSN 0027-8424

Kumar, M.S., Carson, M., Hussain, M.M. & Murthy, H.M. (2002). Structures of Apolipoprotein A-II and a Lipid-Surrogate Complex Provide Insights into Apolipoprotein-Lipid Interactions. *Biochemistry*, Vol. 41, No. 39, (October 2002), pp. 11681-11691, ISSN 0006-2960

Kyte, J. & Doolittle, R.F. (1982). A Simple Method for Displaying the Hydropathic Character of a Protein. *Journal of Molecular Biology*, Vol. 157, No. 1, (May 1982), pp. 105-132, ISSN 0022-2836

Lagerstedt, J.O., Budamagunta, M.S., Oda, M.N. & Voss, J.C. (2007). Electron Paramagnetic Resonance Spectroscopy of Site-Directed Spin Labels Reveals the Structural Heterogeneity in the N-Terminal Domain of ApoA-I in Solution. *The Journal of Biological Chemistry*, Vol. 282, No. 12, (March 2007), pp. 9143-9149, ISSN 0021-9258

Lashuel, H.A. & Lansbury, P.T., Jr. (2006). Are Amyloid Diseases Caused by Protein Aggregates That Mimic Bacterial Pore-Forming Toxins? *Quarterly Reviews of Biophysics*, Vol. 39, No. 2, (May 2006), pp. 167-201, ISSN 0033-5835

Liang, H.Q., Rye, K.A. & Barter, P.J. (1995). Cycling of Apolipoprotein A-I between Lipid-Associated and Lipid-Free Pools. *Biochimica et Biophysica Acta*, Vol. 1257, No. 1, (June 1995), pp. 31-37, ISSN 0006-3002

Mas-Oliva, J., Moreno, A., Ramos, S., Xicohtencatl-Cortes, J., Campos, J. & Castillo, R. (2003). Monolayers of Apolipoprotein AII at the Air/Water Interface, In: *Frontiers in Cardiovascular Health*, Dhalla, N.S., Chockalingam, A., Berkowitz, H.I. & Singal,

P.K., pp. (341-352), Kluwer Academic Publishers, ISBN 978-1-4020-7451-6, Boston, U.S.A.

Meador, W.E., Means, A.R. & Quiocho, F.A. (1992). Target Enzyme Recognition by Calmodulin: 2.4 a Structure of a Calmodulin-Peptide Complex. Vol. 257, No. 5074, (August 1992), pp. 1251-1255, ISSN 0036-8075

Mendoza-Espinosa, P., García-González, V., Moreno, A., Castillo, R. & Mas-Oliva, J. (2009). Disorder-to-Order Conformational Transitions in Protein Structure and Its Relationship to Disease. *Molecular and Cellular Biochemistry*, Vol. 330, No. 1-2, (October 2009), pp. 105-120, ISSN 1573-4919

Mendoza-Espinosa, P., Moreno, A., Castillo, R. & Mas-Oliva, J. (2008). Lipid Dependant Disorder-to-Order Conformational Transitions in Apolipoprotein CI Derived Peptides. *Biochemical and Biophysical Research Communications*, Vol. 365, No. 1, (January 2008), pp. 8-15, ISSN 1090-2104

Mishra, V.K., Palgunachari, M.N., Datta, G., Phillips, M.C., Lund-Katz, S., Adeyeye, S.O., Segrest, J.P. & Anantharamaiah, G.M. (1998). Studies of Synthetic Peptides of Human Apolipoprotein A-I Containing Tandem Amphipathic Alpha-Helixes. *Biochemistry*, Vol. 37, No. 28, (July 1998), pp. 10313-10324, ISSN 0006-2960

Nylander, T. & Wahlgren, M.N. (1997). Forces between Adsorbed Layers of Beta-Casein. *Langmuir: The ACS Journal of Surfaces and Colloids*, Vol. 13, No. 23, (November 1997), pp. 6219-6225, ISSN 0743-7463

Oda, M.N., Forte, T.M., Ryan, R.O. & Voss, J.C. (2003). The C-Terminal Domain of Apolipoprotein A-I Contains a Lipid-Sensitive Conformational Trigger. *Nature Structural Biology*, Vol. 10, No. 6, (June 2003), pp. 455-460, ISSN 1072-8368

Ohnishi, S. & Takano, K. (2004). Amyloid Fibrils from the Viewpoint of Protein Folding. *Cellular and Molecular Life Sciences : CMLS*, Vol. 61, No. 5, (March 2004), pp. 511-524, ISSN 1420-682X

Okon, M., Frank, P.G., Marcel, Y.L. & Cushley, R.J. (2001). Secondary Structure of Human Apolipoprotein A-I(1-186) in Lipid-Mimetic Solution. *Federation of European Biochemical Societies Letters*, Vol. 487, No. 3, (January 2001), pp. 390-396, ISSN 0014-5793

Okon, M., Frank, P.G., Marcel, Y.L. & Cushley, R.J. (2002). Heteronuclear NMR Studies of Human Serum Apolipoprotein A-I. Part I. Secondary Structure in Lipid-Mimetic Solution. *Federation of European Biochemical Societies Letters*, Vol. 517, No. 1-3, (April 2002), pp. 139-143, ISSN 0014-5793

Oram, J.F. (2002). ATP-Binding Cassette Transporter A1 and Cholesterol Trafficking. *Current Opinion in Lipidology*, Vol. 13, No. 4, (August 2002), pp. 373-381, ISSN 0957-9672

Parker, J.L., Christenson, H.K. & Ninham, B.W. (1989). Device for Measuring the Force and Separation between Two Surfaces Down to Molecular Separation. *Review of Scientific Instruments*, Vol. 60, No. 10, (October 1989), pp. 3135-3138, ISSN 0034-6748

Ramos, S., Campos-Teran, J., Mas-Oliva, J., Nylander, T. & Castillo, R. (2008). Forces between Hydrophilic Surfaces Adsorbed with Apolipoprotein AII Alpha Helices. *Langmuir : The ACS Journal of Surfaces and Colloids*, Vol. 24, No. 16, (August 2008), pp. 8568-8575, ISSN 0743-7463

Romero, P., Obradovic, Z. & Dunker, A.K. (2001). Intelligent Data Analysis for Protein Disorder Prediction. *Artificial Intelligence Review*, Vol. 14, No. 6, (December 2000), pp. 447-484, ISSN 0269-2821

Rose, G.D., Fleming, P.J., Banavar, J.R. & Maritan, A. (2006). A Backbone-Based Theory of Protein Folding. *Proceedings of the National Academy of Sciences of the United States of America*, Vol. 103, No. 45, (November 2006), pp. 16623-16633, ISSN 0027-8424

Rozek, A., Buchko, G.W., Kanda, P. & Cushley, R.J. (1997). Conformational Studies of the N-Terminal Lipid-Associating Domain of Human Apolipoprotein C-I by Cd and 1H NMR Spectroscopy. *Protein Science : A Publication of the Protein Society*, Vol. 6, No. 9, (September 1997), pp. 1858-1868, ISSN 0961-8368

Ruíz-García, J., Moreno, A., Brezesinski, G., Möhwald, H., Mas-Oliva, J. & Castillo, R. (2003). Phase Transitions and Conformational Changes in Monolayers of Human Apolipoprotein CI and AII. *Journal of Physical Chemistry B*, Vol. 107, No. 40, (September 2003), pp. 11117–11124, ISSN 1089-5647

Schmidt, H.H., Haas, R.E., Remaley, A., Genschel, J., Strassburg, C.P., Buttner, C. & Manns, M.P. (1997). In Vivo Kinetics as a Sensitive Method for Testing Physiologically Intact Human Recombinant Apolipoprotein A-I: Comparison of Three Different Expression Systems. *Clinica Chimica Acta; International Journal of Clinical Chemistry*, Vol. 268, No. 1-2, (December 1997), pp. 41-60, ISSN 0009-8981

Segrest, J.P., Jones, M.K., De Loof, H., Brouillette, C.G., Venkatachalapathi, Y.V. & Anantharamaiah, G.M. (1992). The Amphipathic Helix in the Exchangeable Apolipoproteins: A Review of Secondary Structure and Function. *Journal of Lipid Research*, Vol. 33, No. 2, (February 1992), pp. 141-166, ISSN 0022-2275

Sickmeier, M., Hamilton, J.A., LeGall, T., Vacic, V., Cortese, M.S., Tantos, A., Szabo, B., Tompa, P., Chen, J., Uversky, V.N., Obradovic, Z. & Dunker, A.K. (2007). Disprot: The Database of Disordered Proteins. *Nucleic Acids Research*, Vol. 35, Suppl. 1, (January 2007), pp. D786-793, ISSN 1362-4962

Su, C.T., Chen, C.Y. & Hsu, C.M. (2007). Ipda: Integrated Protein Disorder Analyzer. *Nucleic Acids Research*, Vol. 35, Suppl. 2, (July 2007), pp. W465-472, ISSN 1362-4962

Swaney, J.B. & Weisgraber, K.H. (1994). Effect of Apolipoprotein C-I Peptides on the Apolipoprotein E Content and Receptor-Binding Properties of Beta-Migrating Very Low Density Lipoproteins. *Journal of Lipid Research*, Vol. 35, No. 1, (January 1994), pp. 134-142, ISSN 0022-2275

Tailleux, A., Duriez, P., Fruchart, J.C. & Clavey, V. (2002). Apolipoprotein A-II, HDL Metabolism and Atherosclerosis. *Atherosclerosis*, Vol. 164, No. 1, (September 2002), pp. 1-13, ISSN 0021-9150

Tamm, L.K. (2003). Hypothesis: Spring-Loaded Boomerang Mechanism of Influenza Hemagglutinin-Mediated Membrane Fusion. *Biochimica et Biophysica Acta*, Vol. 1614, No. 1, (July 2003), pp. 14-23, ISSN 0006-3002

Tanaka, M., Koyama, M., Dhanasekaran, P., Nguyen, D., Nickel, M., Lund-Katz, S., Saito, H. & Phillips, M.C. (2008). Influence of Tertiary Structure Domain Properties on the Functionality of Apolipoprotein A-I. *Biochemistry*, Vol. 47, No. 7, (February 2008), pp. 2172-2180, ISSN 0006-2960

Tartaglia, G.G., Pawar, A.P., Campioni, S., Dobson, C.M., Chiti, F. & Vendruscolo, M. (2008). Prediction of Aggregation-Prone Regions in Structured Proteins. *Journal of Molecular Biology*, Vol. 380, No. 2, (July 2008), pp. 425-436, ISSN 1089-8638

Tompa, P. (2002). Intrinsically Unstructured Proteins. *Trends in Biochemical Sciences*, Vol. 27, No. 10, (October 2002), pp. 527-533, ISSN 0968-0004

Uversky, V.N. (2002). What Does It Mean to Be Natively Unfolded? *European Journal of Biochemistry / Federation of European Biochemical Societies*, Vol. 269, No. 1, (January 2002), pp. 2-12, ISSN 0014-2956

Uversky, V.N., Gillespie, J.R. & Fink, A.L. (2000). Why Are "Natively Unfolded" Proteins Unstructured under Physiologic Conditions? *Proteins*, Vol. 41, No. 3, (November 2000), pp. 415-427, ISSN 0887-3585

Wang, G. (2002). How the Lipid-Free Structure of the N-Terminal Truncated Human ApoA-I Converts to the Lipid-Bound Form: New Insights from NMR and X-Ray Structural Comparison. *Federation of European Biochemical Societies Letters*, Vol. 529, No. 2-3, (October 2002), pp. 157-161, ISSN 0014-5793

Wang, G.,Sparrow, J.T. & Cushley, R.J. (1997). The Helix-Hinge-Helix Structural Motif in Human Apolipoprotein A-I Determined by NMR Spectroscopy. *Biochemistry*, Vol. 36, No. 44, (November 1997), pp. 13657-13666, ISSN 0006-2960

Wang, G., Treleaven, W.D. & Cushley, R.J. (1996). Conformation of Human Serum Apolipoprotein A-I(166-185) in the Presence of Sodium Dodecyl Sulfate or Dodecylphosphocholine by 1H-NMR and CD. Evidence for Specific Peptide-SDS Interactions. *Biochimica et Biophysica Acta*, Vol. 1301, No. 3, (June 1996), pp. 174-184, ISSN 0006-3002

Wei, A., Rubin, H., Cooperman, B.S. & Christianson, D.W. (1994). Crystal Structure of an Uncleaved Serpin Reveals the Conformation of an Inhibitory Reactive Loop. *Nature Structural Biology*, Vol. 1, No. 4, (April 1994), pp. 251-258, ISSN 1072-8368

Weinberg, R.B. & Spector, M.S. (1985). Human Apolipoprotein A-IV: Displacement from the Surface of Triglyceride-Rich Particles by HDL2-Associated C-Apoproteins. *Journal of Lipid Research*, Vol. 26, No. 1, (January 1985), pp. 26-37, ISSN 0022-2275

Westermark, P., Mucchiano, G., Marthin, T., Johnson, K.H. & Sletten, K. (1995). Apolipoprotein A1-Derived Amyloid in Human Aortic Atherosclerotic Plaques. *The American Journal of Pathology*, Vol. 147, No. 5, (November 1995), pp. 1186-1192, ISSN 0002-9440

Wu, Z., Wagner, M.A., Zheng, L., Parks, J.S., Shy, J.M., 3rd, Smith, J.D., Gogonea, V. & Hazen, S.L. (2007). The Refined Structure of Nascent HDL Reveals a Key Functional Domain for Particle Maturation and Dysfunction. *Nature Structural & Molecular Biology*, Vol. 14, No. 9, (September 2007), pp. 861-868, ISSN 1545-9993

Xicohtencatl-Cortes, J.,Castillo, R. & Mas-Oliva, J. (2004b). In Search of New Structural States of Exchangeable Apolipoproteins. *Biochemical and Biophysical Research Communications*, Vol. 324, No. 2, (November 2004), pp. 467-470, ISSN 0006-291X

Xicohtencatl-Cortes, J.,Mas-Oliva, J. & Castillo, R. (2004a). Phase Transitions of Phospholipid Monolayers Penetrated by Apolipoproteins. *Journal of Physical Chemistry B*, Vol. 108, No. 22, (April 2004), pp. 7307-7315, ISSN 1089-5647

Part 2

Others

5

Protein-Protein Interaction Networks: Structures, Evolution, and Application to Drug Design

Takeshi Hase[1,2] and Yoshihito Niimura[3]

[1]Department of Bioinformatics, Graduate School of Biomedical Science,
Tokyo Medical and Dental University, Tokyo,
[2]The Systems Biology Institute, Tokyo,
[3]Department of Bioinformatics, Medical Research Institute,
Tokyo Medical and Dental University, Tokyo,
Japan

1. Introduction

Since proteins exert their functions through interaction with other proteins rather than in isolation, networks of protein interactions are inevitable for understanding protein functions, disease mechanisms, and discovering novel targets of therapeutic drugs (Hase et al. 2009, Barabasi et al. 2011, Vidal et al. 2011). With the recent influx of genome-wide data of protein interactions, many researchers have studied on the structures and statistics of protein-protein interaction networks (PINs). To discover novel drug target genes, it is informative to understand topological and statistical characteristics of PINs, and how disease and drug target genes are distributed over the networks. Moreover, because those statistical properties of PINs are the results of long-term evolution, analysis of the PIN architecture from the viewpoint of comparative genomics and molecular evolution is of particular importance.

In this chapter, we will first summarize our current knowledge of the statistical properties of PINs. We then argue on possible evolutionary mechanisms generating those properties and review the studies related to drug discovery and diseases as an application of the analyses of PIN structure. Finally, we briefly discuss the possibilities of medical studies as an integration of network and evolutionary biology.

2. Genome-wide data of protein-protein interactions

Genome-wide protein-protein interaction data have been obtained from several organisms, including *Escherichia coli* (Arifuzzaman et al. 2006), *Saccharomyces cerevisiae* (Uetz et al. 2000, Ito et al. 2001, Guldener et al. 2006, Reguly et al. 2006, Yu et al. 2008), *Plasmodium falciparum* (LaCount et al. 2005), *Arabidopsis thaliana* (Arabidopsis Interactome Mapping Consortium 2011), *Caenorhabditis elegance* (Li et al. 2004, Simonis et al. 2009), *Drosophila melanogaster* (Giot et al. 2003), and *Homo sapiens* (Rual et al. 2005, Stelzl et al. 2005). Table 1 summarizes the PIN datasets that are currently available. These data were mainly obtained by high-throughput experimental techniques such as yeast two-hybrid (Y2H) screens and tandem affinity purification followed by mass spectrometry (APMS) screens (Deane et al. 2002, Parrish et al. 2006, Lavallee-Adam et al. 2011), as well as extensive literature curation by experts.

Species	Number of proteins	Number of interactions	Data type	References
Mycoplasma pneumoniae	410	1,058	APMS	Kuhner et al. (2009)
MRSA 252	608	13,219	APMS	Cherkasov et al. (2011)
Treponema pallidum	726	3,649	Y2H	Titz et al. (2008)
Mesorhizobium loti	1,804	3,121	Y2H	Shimoda et al. (2008)
Escherichia coli	2,448	8,625	APMS	Arifuzzaman et al. (2006)
Campylobacter jejuni	1,301	11,557	Y2H	Parrish et al. (2007)
Yeast	1,647	2,518	Y2H	Yu et al. (2008)
	3,224	11,291	Literature curated	Reguly et al. (2006)
	3,891	7,270	Manually curated	MIPS
	3,278	4,549	Y2H	Ito et al. (2001)
	1,004	957	Y2H	Uetz et al. (2000)
Malaria parasite	1,267	2,726	Y2H	LaCount et al. (2005)
Arabidopsis thaliana	2,661	5,664	Y2H	Arabidopsis Interactome Mapping Consortium (2011)
Worm	2,898	5,240	Y2H	Li et al. (2004)
	2,528	3,864	Y2H	Simonis et al. (2009)
Fly	4,679	4,780	Y2H	Giot et al. (2003)
	2,477	3,546	Y2H	Pacifico et al. (2006)
Human	2,783	6,007	Y2H, Literature curated	Rual et al. (2005)
	1,613	3,101	Y2H	Stelzl et al. (2005)

Table 1. PIN datasets. Y2H, Yeast two-hybrid screens; APMS, tandem affinity purification followed by mass spectrometry screens. "Manually curated" indicates that interactions obtained from high-throughput screens and literatures are manually integrated by experts.

Y2H screens examine an interaction between two proteins, by expressing these genes in yeast nucleus as fusion proteins (Parrish et al. 2006). One protein is fused to a DNA-binding domain of a transcription factor (*e.g.*, Gal4 and LexA), and the other protein is fused to a transcription activation domain of the transcription factor. When two proteins interact with each other, DNA-binding domain and activation domain are indirectly connected. The activation domain can then interact with the transcription start site of the reporter genes (*e.g.*, LacZ). From the expression of the reporter gene, the interaction between two proteins can be detected. In APMS screens, affinity purification selectively purifies a protein complex that includes a protein of interest (bait protein) (Lavallee-Adam et al. 2011). Then, from the purified complex, mass spectrometry identifies possible interacting partners of the bait protein.

It has been pointed out that genome-wide PIN data identified by high-throughput experiments contains a large number of false positive interactions (Hakes et al. 2008). Y2H screens may detect possible interactions between two proteins that actually reside in different subcellular localizations (Deane et al. 2002). APMS studies identify many false positive interactions caused by inadequate purification (Lavallee-Adam et al. 2011).

Literature-curated PIN datasets are likely to be more reliable, because interactions included in such datasets were obtained from small-scale experiments. However, those data are derived from hypothesis-driven researches focusing on several proteins that are supposed to be biologically important, and thus the datasets can be highly biased (Arabidopsis Interactome Mapping Consortium 2011). Therefore, to study the global structure of PINs, researchers should use several datasets obtained by various methods.

3. Statistical properties of PINs

In PINs, a protein and a physical interaction between two proteins are represented as a node and a link, respectively. A series of studies have revealed that PINs have several interesting properties from the viewpoint of network architecture.

3.1 Scale-freeness

The number of links for a given node is called a degree. The degree distribution $P(k)$, the fraction of nodes with k degrees in a network, has been used to characterize the global structure of a network.

Erdös and Renyi (1960) investigated a random network with N nodes, in which links are attached between each pair of nodes with a uniform probability p. This network contains approximately $pN(N-1)/2$ randomly placed links. Erdös and Renyi (1960) showed that, in a random network, the distribution $P(k)$ follows the Poisson distribution (Fig 1A, left). Therefore, most nodes have degrees that are nearly equal to the mean degree $<k>$ among all nodes in the network.

On the other hand, the distribution $P(k)$ of various technological, social, and biological networks including PINs is known to follow the power law, *i.e.*, $P(k) \sim k^{-\gamma}$ (Albert et al. 1999; Fig 1A, right). These networks are highly heterogeneous; they have a large number of low-degree nodes and a small but significant number of high-degree nodes that are called hubs. A network following the power law does not have a typical degree characterizing most nodes in the network (*e.g.*, the mean degree $<k>$ in a random network), and thus it is called a "scale-free" network. It was shown that scale-free networks are very robust against random removal of nodes, although selective removal of hubs drastically changes their structures (Jeong et al. 2001, Han et al. 2004).

3.2 Small-worldness

The cluster coefficient of nodes i is defined as $C_i = 2e_i/k_i(k_i-1)$, where k_i is the degree of node i, and e_i is the number of links among k_i neighbors of node i (Watts & Strogatz 1998) (see Fig. 1B). In other words, e_i is the number of triangles that pass through node i. C_i is equal to one when all neighbors of node i fully interact with one another, while C_i is 0 when there are no links among the neighbors of node i. The mean of the cluster coefficient among all nodes, $<C>$, reflects the density of triangles ("cliques") within a network.

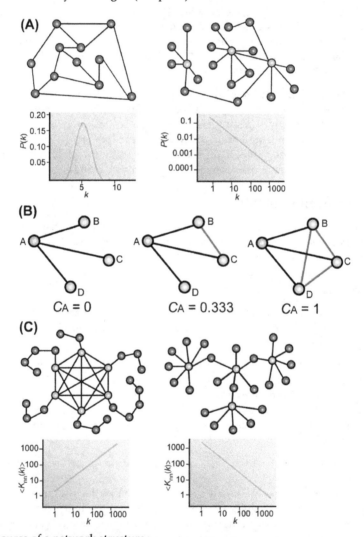

Fig. 1. Measures of a network structure
(A) A random network (left) and a scale-free network (right). The degree distribution $P(k)$ is shown below the networks. (B) Cluster coefficient. Red lines represent links among three neighbors of node A. The numbers of links (e_A) among nodes B, C, and D (the neighbors of

node A) in the left, middle, and right networks are 0, 1, and 3, respectively. The cluster coefficient C_A of node A is shown for each network. **(C)** Assortative (left) and disassortative (right) networks. The distribution of $<K_{nn}(k)>$ is shown below the networks. Blue and Red nodes indicates hubs and non-hubs, respectively.

The shortest path length between a pair of nodes is the smallest number of links (distance) that are necessary for travelling from one node to the other (Barabasi & Oltvai 2004). The mean shortest path length among all possible pairs of nodes in a network is denoted by $<L>$. Watts and Strogatz (1998) found that a random network has a much smaller value of $<L>$ compared with a regular lattice. Based on this observation, they defined a "small-world" network as a network that has a value of $<L>$ as small as a random network but is highly clustered like a regular lattice. In a random network, $<L> \sim \log N / \log <k>$, and $<C> = <k>/N$, where N is the number of nodes.

In PINs, the value of $<L>$ is small and the value of $<C>$ is much higher than a random network; therefore, PINs are generally considered to be small-world networks. However, several studies showed that PINs are actually "ultra-small", because $<L>$ is considerably smaller than that in a random network (Chung & Lu 2002, Cohen & Havlin 2003, Hase et al. 2008). In a PIN, proteins are located close to each other, suggesting that perturbations given to a single protein would affect the behaviour of many other proteins and even the entire PIN.

3.3 Assortativity

Another statistic characteristic of a network is the correlation between degrees of nodes that are linked to each other (Callaway et al. 2001, Newman 2002, Costa et al. 2007). Pearson correlation coefficient r of the degrees at both ends of a link is used to evaluate the degree correlation. Networks with $r > 0$ and $r < 0$ are called as assortative and disassortative networks, respectively. In an assortative network, hubs tend to be connected to each other (Fig 1C, left), while in a disassortative network, hubs tend to have links to low-degree nodes (Fig 1C, right).

$<K_{nn}(k)>$, the mean degree among the neighbors of all k-degree nodes ("nn" in $<K_{nn}(k)>$ means "nearest neighbors"), is also used to evaluate the assortativity of a network (Pastor-Satorras et al. 2001, Maslov & Sneppen 2002, Costa et al. 2007, Hase et al. 2008). In an assortative network, $<K_{nn}(k)>$ increases as k increases, while $<K_{nn}(k)>$ in a disassortative network follows decreasing functions of k (Fig 1C). If there are no correlations between degrees of nodes at both ends of a link (e.g., $r = 0$), $<K_{nn}(k)>$ is independent from k and is equal to $<k^2>/<k>$.

It has been shown that the yeast PIN is a disassortative network (Maslov & Sneppen 2002). Therefore, in the yeast PIN, interactions between high- and low-degree nodes are favoured, while those between hubs are suppressed. The biological significance of this structure is unclear. Maslov and Sneppen (2002) proposed that, in the yeast PIN, a hub protein forms a functional module of a cell together with a large number of low-degree neighbors. They then hypothesized that the suppression of links between hubs minimizes unfavourable cross-talks among different functional modules and makes networks robust against perturbations.

If this hypothesis is true, disassortative structure observed in the yeast PIN is under the natural selection, and the disassortativity should be commonly found among PINs in any organisms. However, by examining PINs from five eukaryote species, Hase et al. (2010) found that the disassortative structure is not a common feature of PINs. The distribution of $<K_{nn}(k)>$ in the PIN can be approximated by $<K_{nn}(k)> \sim k^{-\nu}$, and the value of ν is used to quantify the extent of disassortative structure of a network. Hase et al. (2010) showed that the yeast, worm, fly, and human PINs are disassortative (ν = 0.47, 0.29, 0.35, and 0.26, respectively), while the malaria parasite PIN is not disassortative (ν = 0.02). This observation indicates that the "selectionist view" by Maslov and Sneppen (2002) is not necessary for explaining the disassortative structure of PINs. In section 4, we will see the evolutionary mechanisms generating the difference in assortativity among species.

4. Evolutionary mechanisms generating structures of PINs

To account for the emergence of PIN architecture mentioned above, researchers developed several network growth models and conducted simulation studies using these models. Moreover, statistical properties of PINs were analyzed from the viewpoint of comparative genomics and molecular evolution. In this section, we review evolutionary studies of PINs.

4.1 Preferential attachment and gene duplication

Barabasi and Albert (1999) suggested that the emergence of scale-freeness can be explained by two basic mechanisms: network growth and preferential attachment. The process of network growth adds a new node into a network (red node in Fig 2A). The process of preferential attachment introduces a new link between the new node and each of the other nodes with the probability proportional to the degree of the latter node. For example, the probability that the red node in Fig. 2A gains a new link connected to a blue node is three times higher than that to a black node (Fig 2A). Due to these two processes, a node with a higher degree gains a larger number of links, and thus the degrees of high-degree nodes increase faster than those of low-degree nodes, generating a scale-free network.

In fact, Eisenberg and Levanon (2003) demonstrated that the number of interactions that a protein gained during its evolution is roughly proportional to the degree of the protein by comparing the genomes of *E. coli*, *A. thaliana*, *Schizosaccharomyces pombe*, and *S. cerevisiae*. This observation is consistent with the preferential attachment.

What is the genetic mechanism of network growth and preferential attachment in the evolution of PINs? A plausible mechanism is gene duplication. Let us consider a small PIN containing both high- (node A) and low-degree nodes (node B, C, and D) (Fig 2B, middle). We assume that the number of nodes in a network increases by gene duplication, and a new node has the same interacting partners as the original node. When node B is duplicated, for example, node A acquires a new link and thus the degree of node A increases by one. When node C or node D is duplicated, the same thing happens. On the other hand, if node A is duplicated, each of the degrees of nodes B, C, and D increases by one. Under the assumption that gene duplication occurs randomly with an equal probability for all nodes, the probability that node A acquires a new link is three times higher than the other node does. In general, when we compare a high-degree node (*e.g.*, A) and a low-degree node (*e.g.*, B), a given node (*e.g*, C) is more likely to be a neighbor of a high-degree node than that of a low-degree node. Therefore, a high-degree node gains new links faster than a low-degree node does. For this reason, gene duplication can account for the mechanism of "rich-get-richer".

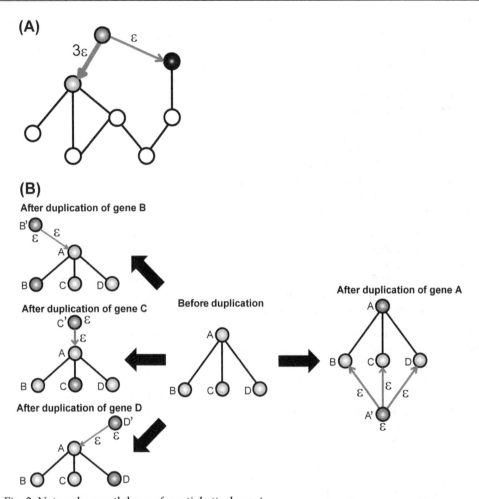

Fig. 2. Network growth by preferential attachment
(A) Preferential attachment. A red node is added to the network. The probability that a new link is attached between red and blue nodes (3ε) is three times higher than that between red and black nodes (ε). (B) Network growth with gene duplication. Red nodes represent duplicated nodes. Gene duplication occurs with an equal probability (ε) for all nodes. When node A is duplicated, degrees of nodes B, C, and D increase by one (right), whereas when either node B, C, or D is duplicated, degree of node A increases by one (left).

4.2 Duplication and divergence model

A pair of genes generated by duplication will undergo one of three fates, namely, (i) neofunctionalization, (ii) subfunctionalization, and (iii) nonfunctionalization. After gene duplication, one of the duplicated genes becomes free from selective pressure because of the presence of redundant copies of the gene. Therefore, the gene can tolerate to the accumulation of random mutations and in some cases acquire a novel function (Ohno 1970).

This process is called neofunctionalization. On the other hand, in subfunctionalization process, each of the duplicated genes accumulates mutations, and the functions of the ancestral gene are assigned to the two genes (Force et al. 1999). In nonfunctionalization process, one of the duplicated genes loses its function and becomes a pseudogene due to deleterious mutations. Among the three processes, neofunctionalization and subfunctionalization contribute to the evolution of proteins (Lynch et al. 2000, Blanc et al. 2004, He et al. 2005, Freilich et al. 2006).

In the duplication-divergence model, neofunctionalization and subfunctionalization are modelled as attachment of new links and removal of the links generated by gene duplication, respectively. As for subfunctionalization process, there are two different models, the symmetric divergence and asymmetric divergence. In the former, links are eliminated from both of the duplicated nodes, while in the latter, elimination of links occurs only in one of the two nodes generated by duplication (Fig 3A).

Wagner (2002) reported that one of the duplicated proteins retain a significantly larger number of interactions than the other. For this reason, several network growth models adopted the asymmetric divergence model (Kim et al. 2002, Wagner 2003, Chung et al. 2003, Ispolatov et al. 2005c). However, "complete" asymmetric divergence in which links are eliminated from only one of the duplicates is unrealistic, and the actual divergence process should be intermediate between symmetric and asymmetric divergence (Hase et al. 2010).

Sole et al. (2002) proposed a model on the basis of neofunctionalization and asymmetric divergence. According to their model, after duplication generates a new node, neofunctionalization process attaches a new link between either of the duplicated nodes and each of the other nodes with a uniform probability θ, and then asymmetric divergence eliminates links to only one of the duplicated nodes with a uniform probability α (Fig 3A). Simulation and analytical studies have demonstrated that this model can generate scale-free networks with a small-world property (Sole et al. 2002, Kim et al. 2002, Pastor-Satorras et al. 2003, Chung et al. 2003, Raval 2003).

However, it has been pointed out that some statistical features of PINs could not be regenerated by the model of Sole et al. (2002). The yeast and fly PINs show a much larger $<C>$ than the networks by Sole et al. with the same number of nodes and links as the actual PINs (Sole et al. 2002, Middendorf et al. 2005, Ispolatov et al. 2005a). To overcome this problem, Vazquez et al. (2003) proposed the heterodimerization (HD) model. In their model, symmetric divergence eliminates links from both of the duplicated nodes with a uniform probability α, and the HD process attaches a new link between two duplicated nodes with another uniform probability β, forming a heterodimer (Fig 3A).

When gene duplication occurs for a self-interacting protein, the duplicated proteins will interact to each other. Therefore, β in Vazquez et al. (2003) represents the probability that a randomly selected protein is self-interacting and the new HD link between two duplicated proteins survives after divergence. Simulation and analytical studies have showed that the HD model could reproduce scale-free networks with a similar $<C>$ to the yeast and fly PINs (Vazquez et al. 2003, Middendorf et al. 2005, Ispolatov et al. 2005a). This is because an HD process creates triangles, and a network containing a large number of triangles shows a high value of $<C>$. A computational study based on machine learning technique showed that the

HD model could best reproduce the fly PIN among seven network growth models (Middendorf et al. 2005).

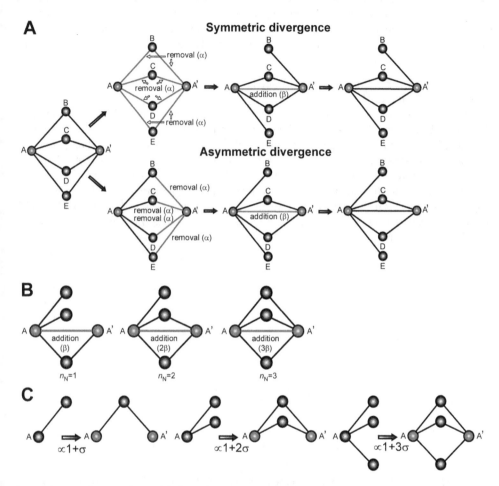

Fig. 3. Network growth models based on gene duplication and divergence
A pair of two red nodes are generated by gene duplication. **(A)** The HD model with asymmetric or symmetric divergence processes. Nodes A and A′ are generated by gene duplication. In the symmetric divergence, each of the links to nodes A and A′ is eliminated with a uniform probability α. On the other hand, in the asymmetric divergence, each of the links to node A′ is eliminated with a uniform probability α. After the divergence process, an HD link (a red line) between two duplicated nodes (nodes A and A′) is attached with a uniform probability β. **(B)** The NHD model. An HD link (red link) is attached between nodes A and A′ with a probability proportional to the number (n_N) of common neighbors shared by these nodes. **(C)** The DDD model. A probability of duplication of a given node is dependent on the degree of the node. If a node has k links, the node is duplicated with the probability proportional to $1 + k\sigma$, where σ is a parameter of the duplicability of a node.

4.3 Non-uniform heterodimerization model

By conducting simulation studies, Hase et al. (2008) showed that, to reproduce the value of <C> in the yeast PIN by the HD model, the number of HD links in the networks by the HD model has to be much larger than that in the yeast PIN. Similar observation was made for the fly PIN (Ispolatov et al. 2005a and b). This means that the HD model is insufficient for explaining the evolution of PINs.

As shown in Fig. 3B, when two duplicated nodes share one, two, and three common neighbors, an HD link between them generates one, two, and three new triangles, respectively. The high <C> in a PIN indicates that the network contains many triangles. Therefore, if a new HD link is attached more preferentially between duplicated nodes sharing a larger number of common neighbors, the value of <C> in a simulation-generated network is expected to become higher. By considering in this way, Hase et al. (2008) proposed the non-uniform heterodimerization (NHD) model in which a new HD link is added between duplicated nodes with a probability proportional to the number of neighbors shared by those nodes (Fig 3B). Simulation studies demonstrated that the NHD model could indeed reproduce both the high value of <C> and the small number of HD links in the yeast PIN.

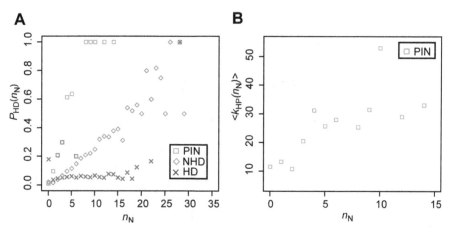

Fig. 4. HD links in the yeast PIN and in the networks by the HD and NHD models (Hase et al. 2008). **(A)** Distribution of $P_{HD}(n_N)$, the probability that an HD link exists between two homologous proteins when they share n_N common neighbors. Green squares, blue diamonds, and red crosses indicate the values for the yeast PIN, the network by the NHD model, and that by the HD model, respectively. **(B)** Distribution of $k_{HP}(n_N)$, the mean degree of proteins that are connected by an HD link and share n_N common neighbors.

In the evolution of PINs, duplication of a self-interacting protein adds an HD link between duplicated proteins. Some HD links were conserved in evolution, while others were eliminated because of occurrence of mutations at interacting sites in these duplicated proteins. In the HD model, the survival rate of HD links is uniform; on the other hand, the NHD model assumes it to be proportional to the number of their common neighbors (Fig. 4A). In the yeast PIN, the probability that two homologous node retain an HD link increases as the number of neighbors shared by the two nodes increases, which is consistent to the NHD model rather than the HD model (Fig. 4A).

A possible explanation for this observation is as follows. It is expected that, when a given pair of proteins share a large number of common neighbors, the degree of these proteins should be high. In fact, in the yeast PIN, when two homologous proteins are connected by a HD link, there is a positive correlation between the number of common neighbors to the homologues and the mean degree of the two proteins (Fig. 4B). Moreover, several studies showed that high-degree proteins tend to show low evolutionary rate in the yeast PIN (Fraser et al. 2002, 2003, Fraser 2005). Therefore, it is suggested that the survival rates of HD links are also positively correlated with the number of common neighbors shared by the two homologous proteins.

4.4 Degree-dependent duplicability and assortativity

Duplication and divergence models including the NHD model can explain various aspects of the architecture of PINs (Pastor-Satrras et al. 2003, Vazquez 2003, Hase et al. 2008). However, these models cannot explain the differences in overall structures of PINs among species. As mentioned in section 3, the yeast, worm, fly, and human PINs are disassortative, while the malaria parasite PIN is non-disassortative.

A possible evolutionary scenario that can explain the difference in assortativity of PINs among different species is as follows (Hase et al. 2010). Let us consider a disassortative network containing low- and high-degree nodes (*e.g.*, A and C, respectively), in which the low- and high-degree nodes are linked to each other (Fig 5A, middle). Duplication of a low-degree node (*e.g.*, node A) causes the value of ν in the disassortative network to be higher, because the degree of its high-degree neighbor increases (Fig 5A, left). On the other hand, duplication of a high-degree node (*e.g.*, node C) makes the degree of its low-degree neighbors higher, and thus the value of ν decreases (Fig 5A, right). For this reason, duplication of low- and high-degree nodes would make the value of ν in a disassortative network larger and smaller, respectively.

Hase et al. (2010) proposed a novel duplication and divergence model named "degree-dependent duplication (DDD) model", in which duplication of nodes occurs depending on their degree (see Fig 3C). Simulation studies based on the DDD model revealed that preferential duplication of low-degree nodes can successfully reproduce the disassortative structure observed in the yeast, worm, and fly PINs, while preferential duplication of high-degree nodes generate non-disassortative networks similar to the malaria parasite PIN (see Fig 5B and 5C). Moreover, Hase et al. (2010) evaluated the dependency of gene duplicability on their degrees by analyzing orthologous relationships of genes extracted from 55 eukaryotic proteomes. The analyses demonstrated that proteins with a lower degree indeed have higher duplicability in disassortative PINs (the yeast, worm, and fly PINs) (Fig 5D), whereas high-degree proteins tend to have high duplicability in non-disassortative PINs (the malaria parasite PIN) (Fig 5E). Therefore, it is suggested that assortativity of a PIN is related with the gene duplicability dependent on the degrees of genes. If this is the case, disassortative structure of PINs is merely a byproduct of preferential duplication of low-degree proteins, and we do not need to assume any adaptive meaning for this structure, as mentioned in section 3.

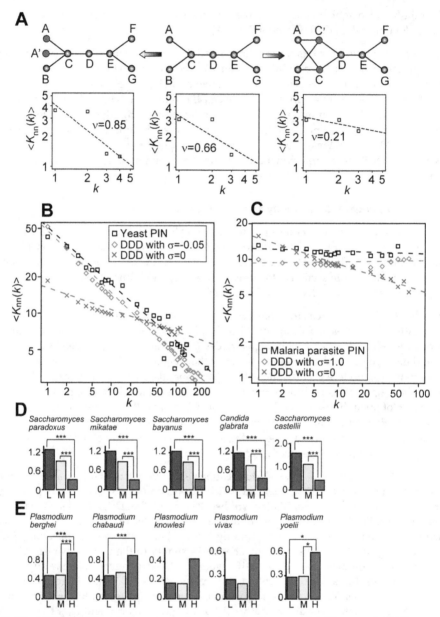

Fig. 5. The DDD model and the extent of assortativity in networks (Hase et al. 2010).
(A) Duplication of a node alters the distribution of $<K_{nn}(k)>$ and the value of v in a network.
A diagram below a network shows the distribution of $<K_{nn}(k)>$ and the value of v in the
network. (B) The distribution of $<K_{nn}(k)>$ in the networks generated by the DDD model for
yeast. Blue diamonds and red crosses show the results of simulation with σ = -0.05 and 0,
respectively (as for σ, see Fig. 3C). Black squares represent $<K_{nn}(k)>$ in the yeast PIN. Dashed

lines in black, blue, and red represent $k^{-0.47}$, $k^{-0.51}$, and $k^{-0.18}$, respectively. **(C)** The distribution of $<K_{nn}(k)>$ in the networks generated by the DDD model for malaria parasite. Blue diamonds and red crosses show the results of simulation with $\sigma = 1.0$ and 0, respectively. Black squares represent $<K_{nn}(k)>$ in the malaria parasite PIN. Dashed lines in black, blue, and red represent $k^{-0.02}$, $k^{0.01}$, and $k^{-0.22}$, respectively. **(D)** and **(E)** indicate correlations between the degree and the duplicability in the yeast and malaria parasite PINs, respectively. Bars in blue, yellow, and red show the mean duplicability among low-, middle-, and high-degree proteins, respectively. A species name above each diagram denotes the species of which genome was compared with *S. cerevisiae* or *P. falciparum*. *, **, and *** represent $P < 0.05$, $P < 0.01$, and $P < 0.001$, respectively, by the Wilcoxon rank-sum test with the Bonferroni correction.

5. Structures of PINs and their relationships with disease genes and drug targets

As we have seen above, PINs are characterized by several interesting properties that are different from those of a random network. Therefore, understanding diseases and mechanisms of drug action in the context of PIN architecture may allow us to address some fundamental properties of disease genes and drug target molecules. Indeed, number of disease genes and that of drug targets are very small. Only 10% of the human genes are known to be disease genes (Amberger et al. 2009), and only 435 genes are target genes of therapeutic drugs (Rask-Andersen et al. 2011). Why is the number of drug targets and disease genes so small? Are they distributed randomly over the human PIN? Are there any quantifiable correlations between drug target genes and their statistical properties in the human PIN? To address these questions, drug target and disease genes were mapped onto the human PIN and their statistical properties in the PIN were investigated. Moreover, by using biological networks including the human PIN data, several studies showed that side effects of drugs depend on their statistical features in the network. In this and subsequent sections, we review the application of network analyses to medical researches.

5.1 Statistical properties of disease genes and drug targets in the human PIN

Elimination of a hub protein affects many proteins in a network (Jeong et al. 2001, Yu et al. 2008). Therefore, it was previously hypothesized that genes encoding hub proteins are associated with diseases (Barabasi et al. 2011). Several studies reported that the mean degree among disease genes is in fact significantly higher than that among non-disease genes (Wachi et al. 2005, Jonsson & Bates 2006, Xu & Li 2006).

A human gene is defined to be essential, when knock-out of its orthologous gene causes embryonic and postnatal lethality or sterility in mouse (Liang & Li 2007). Liang & Li (2007) reported that essential genes tend to encode hub proteins in the human PIN.

However, Wachi et al. (2005), Jonsson & Bates (2006), and Xu & Li (2006) took no account for the fact that there are only a small number of disease genes that are also essential (essential disease genes), while vast majority of disease genes are actually non-essential. Because essential disease genes encode hub proteins, the mean degree of disease genes became apparently high in the three studies. In contrast, non-essential disease genes do not show any tendency to encode hub proteins (Goh et al. 2007). Rather, they tend to encode low- and middle-degree proteins (Feldman et al. 2008). Mutations in high-degree proteins cause dysfunctionality of many neighbor proteins, leading severe impairment of developmental and physiological processes. Individuals having such mutations cannot survive until reproductive years and are likely to be removed from population. For this reason, non-essential disease genes are enriched among low- and middle-degree genes.

Hase et al. (2009) investigated drug target genes to see whether they have specific statistical features in the PIN or not. They found that most drug target genes are middle-degree proteins and some are low-degree, while there are almost no drug targets among high-degree proteins (see Fig 6). The degree distribution is similar to that of disease genes, and, not surprisingly, drug target genes significantly overlap with disease genes (Yao & Rzhetsky 2008). These results indicate that middle-degree proteins are likely to be most advantageous targets for therapeutic drugs.

Oncogenes tend to be high-degree proteins (Jonsson & Bates 2006), and thus they are less likely to be targets for drugs, or one must accept major potential side effects. A possible strategy for designing anti-cancer therapy with less severe side effects is to develop a novel combination of drug compounds that targets several low- and middle-degree proteins, because such combination could generate synergetic effects to cancer cure, and low- or middle-degree targets are expected to induce less severe side effects compared with high-degree targets.

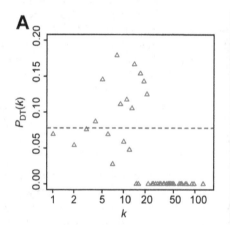 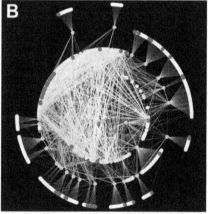

Fig. 6. Degree distribution of drug targets (Hase et al. 2009).
(A)$P_{DT}(k)$ represents the fraction of drug targets to all proteins for the degree of k. The dashed line in red represents the probability that a randomly selected protein is a drug target. **(B)** White, yellow, and blue nodes represent low- ($k = 1 - 5$), middle- ($k = 6 - 30$) and high-degree ($k > 30$) proteins, respectively. Drug targets (red nodes) are mapped on the network. White, yellow, green, and blue links represent interactions between high- and low-degree proteins, those between middle-degree proteins, those between high- and middle-degree proteins, and those between high-degree proteins, respectively. Middle-degree proteins are extensively connected to each other, while links between high-degree proteins are rather suppressed. For clarity, low- and middle-degree proteins that do not have any interactions with high-degree proteins are not shown.

5.2 Predicting candidate drug targets and their side effects based on biological networks

To develop a new drug, it is critical to accurately predict its side effect, because almost 30% of candidate drugs are rejected in clinical stages due to their unexpected toxicity or concerns about drug safety (Kola & Landis 2004, Billingsley 2008). Severe adverse reactions may be found long after the approval of drugs (*e.g.*, Rosiglitazone), and in such cases, those drugs would go out of production (*e.g.*, Rofecoxib) (Moore et al. 2007).

The chemical structures of drugs have been used to predict their adverse side effects and target proteins (Kuhn et al. 2008, Campillos et al. 2008, Yamanishi et al. 2010). Campillos et al. (2008) developed a large-scale database of adverse side effects of drugs. By using the database with information of chemical structure of drugs, they made a similarity metric between two drugs. Under the assumption that drugs with higher similarity in their metric more tend to share the same target proteins, they inferred candidate targets for the drugs.

However, if target proteins of two drugs are close in a molecular network, such drugs may cause similar downstream effects in the network and thus have similar side effects. To understand the molecular mechanisms of drug action and associated adverse effects in greater details, it makes sense to view targets of drugs in the context of biological networks including the genome-wide human interactome (Pache et al. 2008, Zanzoni et al. 2009).

Recently, Brouwers et al. (2011) investigated how side effect similarities of targets depend on their closeness in the human PIN. They found that a certain number of pairs of two drugs without common targets show similar side effects, when they are close in the human PIN. Moreover, Wang et al. (2011) reported that drug side effects are significantly associated with network distances between drug target genes and diseases genes, *i.e.*, targets for failure drugs that make severe adverse side effects are closer to disease genes than targets for approved drugs. Thus, selecting targets that are too close to diseases genes are not always the best strategy (Wang et al. 2011), although the pharmaceutical industry tends to select targets of new drugs that are close with the corresponding disease genes in the biological networks, especially after 1996 (Yildirim et al. 2007).

With recent influx of information of biological networks, especially the human interactome, analyses like Brouwers et al. (2011) or Wang et al. (2011) can be refined and adapted to infer still unknown adverse side effects of drugs and to predict possible target genes. Indeed, by integrating information of the human PIN with similarities between two genes (*e.g.*, GO semantic and sequence similarity) and those between two drugs (*e.g.*, chemical and drug therapeutic similarity), several recent researches attempted to develop a method to predict possible targets for therapeutic drugs (Zhao & Li 2010, Perlman et al. 2011).

6. Possibilities of medical studies with integration of PINs and evolutionary studies

The human PIN is still incomplete and there are many proteins without any information of protein-protein interactions (Venkatesan et al. 2009). Evolutionary information (*e.g.*, evolutionary rate and duplicability) of genes is significantly correlated with their statistical properties in PINs (see sections 2 and 3); therefore, such information can be utilized to complement to the incompleteness of the human PIN.

Rambaldi et al. (2008) reported that most of the cancer genes are singletons and have interactions with many genes. This finding indicates that both gene duplicability and network information are useful for predicting candidate cancer genes. Modification of currently available methods by integrating evolutionary information would improve the accuracy of predicting disease and drug target genes.

Recently, large-scale PINs became available from several prokaryotes, including Methicillin-resistant *Staphylococcus aureus* (MRSA) (Cherkasov et al. 2010), *Treponema pallidum* (Titz et al. 2008), *Campylobacter jejuni* (Parrish et al. 2007), *Mycoplasma pneumonia* (Kuhner et al. 2009), and *Mesorhizobium loti* (Shimoda et al. 2008) (see Table 1). Some of them are pathogenic. By investigating the evolution of their PINs, we may be able to understand the process of acquiring the pathogenicity and developing drug resistance from the viewpoint of network architecture.

Cherkasov et al. (2010) suggested that, in the MRSA PIN, hubs are essential for network stability and may be prospective antimicrobial drug targets. However, almost all known antimicrobial targets have relatively few interactions and hubs have largely been overlooked as drug targets. If hubs in pathogens have no orthologous genes in human and evolve very slowly, by targeting such hubs, we may be able to develop novel antibacterial drugs with high efficacy and small side effects, and without development of resistance to the drugs. With a recent influx of PINs from pathogenic organisms and genomes from various bacterial species, analyses integrating comparative genomics with PINs will become keys to identify still unknown disease mechanisms and novel targets for antibacterial drugs.

7. Conclusion

In this chapter, we describe various aspects of architecture of PINs, such as scale-freeness, small-world properties, and assortativity. Computational studies based on network-growth models and comparative genomics revealed how accumulation of local changes in PINs affects their overall architecture during evolution. We also discussed possible application of PINs and evolutionary studies to medical researches. With expected explosion of OMICs data (*e.g.*, PINs and SNPs from human) in the near future, an integration of networks and genetics will be among the most powerful strategies to elucidate unknown mechanisms of disorders and discover novel targets for efficacious drugs.

8. Acknowledgements

This study was supported by the Ministry of Education, Culture, Sports, Science and Technology, Japan, grant 23770271 to YN.

9. References

Albert, R.; Jeong, H. & Barabasi, AL. (1999). Diameter of the World-Wide Web, *Nature*, vol. 401, pp. 130–131.

Amberger, J.; Bocchini, CA.; Scott, AF. & Harmosh, A. (2009). McKusick's Online Mendelian Inheritance in Man (OMIM), *Nucleic Acids Research*, vol. 37, pp. D793–D796.

Arabidopsis Interactome Mapping Consortium (2011) Evidence for network evolution in an Arabidopsis interactome map, *Science*, vol. 333, pp. 601–607.

Arifuzzaman, M.; Maeda, M.; Itoh, A.; Nishikata, K.; Takita, C.; Saito, R.; Ara, T.; Nakahigashi, K.; Huang, HC.; Hirai, A.; Tsuzuki, K.; Nakamura, S.; Altaf-Ul-Amin, M.; Oshima, T.; Baba, T.; Yamamoto, N.; Kawamura, T.; Ioka-Nakamichi, T.; Kitagawa, M.; Tomita, M.; Kanaya, S.; Wada, C. & Mori, H. (2007). Large-scale identification of protein-protein interaction of Escherichia coli K-12, *Genome Research*, vol. 16, pp. 686–691.

Barabasi, AL. & Albert, R. (1999). Emergence of scaling in random networks, *Science*, vol. 286, pp. 509–512.

Barabasi, AL. & Oltvai, ZN. (2004). Network biology : understanding the cell's functional organization, *Nature Reviews Genetics*, vol. 5, pp. 101–113.

Barabasi, AL.; Gulbahce, N. & Lascalzo, J. (2011). Network medicine : a network-based approach to human disease, *Nature Reviews Genetics*, vol. 12, pp. 56–68.

Billingsley, ML. (2008). Druggable targets and targeted drugs: enhancing the development of new therapeutics, *Pharmacology* vol. 82, pp. 239–244.

Blanc, G. & Wolfe, KH. (2004). Functional divergence of duplicated genes formed by polyploidy during Arabidopsis evolution, *The Plant Cell*, vol. 16, pp. 1679–1691.

Brouwers, L.; Iskar, M.; Zeller, G.; Noort, V. & Bork, P. (2011). Network neighbors of drug targets contribute to drug side effect similarity, *PLoS ONE*, vol. 6, e22187.

Callaway, DS.; Hopcroft, JE.; Kleinberg, JM.; Newman, MEJ. & Strogatz, SH. (2001). Are randomly grown graphs really random?, *Physical Review E*, vol. 64, 041902.

Campillos, M.; Kuhn, M.; Gavin, AC.; Jensen, LJ. & Bork, P. (2008). Drug target identification using side effect similarity, *Science*, vol. 321, pp. 263–266.

Cherkasov, A.; Hsing, M.; Zoraghi, R.; Foster, LJ.; See, RH.; Stoynov, N.; Jiang, J.; Kaur, S.; Lian, T.; Jackson, L.; Gong, H.; Swayze, R.; Amandoron, E.; Hormozdiari, F.; Dao, P.; Sahinalp, C.; Santos-Filho, O.; Axerio-Cilies, P.; Byler, K.; McMaster, WR.; Brunham, RC.; Finlay, BB. & Reiner, NE. (2011). Mapping the protein interaction network in Methicillin-resistant Staphylococcus aureus, *Journal of Proteome Research*, vol. 10, pp. 1139–1150.

Chung, F. & Lu, L. (2002). The average distances in random graphs with given expected degrees, *Proceedings of National Academy of Sciences USA*, vol. 99, pp. 15879–15882.

Chung, F.; Lu, L.; Dewey, TG. & Galas DJ. (2003). Duplication models for biological networks, *Journal of computational biology*, vol. 10, pp. 677–687.

Cohen, R. & Havlin, S. (2003). Scale-free networks are ultrasmall, *Physical Review Letters*, vol. 90, 058701.

Costa, LF.; Rodrigues, FA.; Travieso, G. & Boas, RRV. (2007). Characterization of complex networks : A survey of mesurements, *Advances in Physics*, vol. 56, pp. 167–242.

Deane CM.; Salwinski L.; Xenarios I. & Eisenberg D. (2002) Protein interactions, *Molecular & Cellular Proteomics*, vol. 1, pp. 349–356.

Eisenberg, E. & Levanon, EY. (2003). Preferential attachment in the protein network evolution, *Physical Review Letters*, vol. 91, 138701.

Erdös, P. & Renyi, A. (1960). On the evolution of random graphs, *Publication of the Mathematical Institute of the Hungarian Academy of Science*, vol. 5, pp. 17–61.

Feldman, I.; Rzhetsky, A. & Vitkup, D. (2008). Network properties of genes harbouring inherited disease mutations, *Proceedings of National Academy of Sciences USA*, vol. 105, pp. 4323–4328.

Force, A.; Lynch, M.; Pickett, FB.; Amores, A.; Yan, YL. & Postlethwait, J. (1999). Preservation of duplicate genes by complementary, degenerative mutations, *Genetics*, vol. 151, pp. 1531–1545.

Fraser, HB.; Hirsh, AE.; Steinmetz, LM.; Scharfe, C. & Feldman, MW. (2002). Evolutionary rate in the protein interaction network, *Science*, vol. 296, pp. 750–752.

Fraser, HB.; Wall, DP. & Hirsh, AE. (2003). A simple dependence between protein evolution rate and the number of protein-protein interactions, *BMC Evolutionary Biology*, vol. 3, 11.

Fraser, HB. (2005). Modularity and evolutionary constraint on proteins, *Nature Genetics*, vol. 37, pp. 351–352.

Freilich, S.; Massingham, T.; Blanc, E.; Goldovsky, L. & Thornton, JM. (2006). Relating tissue specialization to the differentiation of expression of singleton and duplicate mouse protein, *Genome Biology*, vol. 7, R89.

Giot, L.; Bader, JS.; Brouwer, C.; Chaudhuri, A.; Kuang, B.; Li, Y.; Hao, YL.; Ooi, CE.; Godwin, B.; Vitols, E.; Vijayadamodar, G.; Pochart, P.; Machineni, H.; Welsh, M.; Kong, Y.; Zerhusen, B.; Malcolm, R.; Varrone, Z.; Collis, A.; Minto, M.; Burgess, S.; McDaniel, L.; Stimpson, E.; Spriggs, F.; Williams, J.; Neurath, K.; Ioime, N.; Agee, M.; Voss, E.; Furtak, K.; Renzulli, R.; Aanensen, N.; Carrolla, S.; Bickelhaupt, E.; Lazovatsky, Y.; DaSilva, A.; Zhong, J.; Stanyon, CA.; Finley, RL Jr.; White, KP.; Braverman, M.; Jarvie, T.; Gold, S.; Leach, M.; Knight, J.; Shimkets, RA.; McKenna, MP.; Chant, J. & Rothberg, JM. (2003). A protein interaction map of Drosophila melanogaster, *Science*, vol. 302, pp. 1727–1736.

Goh, KI.; Cusick, ME.; Valle, D.; Childs, B.; Vidal, M. & Barabasi, AL. (2007). The human disease network, *Proceedings of National Academy of Sciences USA*, vol. 104, pp. 8685–8690.

Guldener, U.; Munsterkotter, M.; Oesterheld, M.; Pagel, P.; Ruepp, A.; Mewes, HW. & Stumpflen, V. (2006). Mpact: the MIPS protein interaction resource on yeast, *Nucleic Acids Research*, vol. 34, pp. D436–441.

Hakes, L.; Pinney, JW.; Robertson, DL. & Lovell, SC. (2008). Protein-protein interaction networks and biology–what's the connection?, *Nature Biotechnology*, vol. 26, pp. 69–72.

Han, JDJ.; Bertin, N.; Hao, T.; Goldberg, DS.; Berriz, GF.; Zhang, LV.; Dupuy, D.; Walhout, AJM.; Cusick, ME.; Roth, FP. & Vidal, M. (2004). Evidence for dynamically organized modularity in the yeast protein-protein interaction network, *Nature*, vol. 430, pp. 88–93.

Hase, T.; Niimura, Y.; Kaminuma, T. & Tanaka, H. (2008). Non-uniform survival rate of heterodimerization links in the evolution of the yeast protein-protein interaction network, *PLoS ONE*, vol. 3, e1667.

Hase, T.; Tanaka, H.; Suzuki, Y.; Nakagawa, S. & Kitano, H. (2009). Structure of protein interaction networks and their implications on drug design, *PLoS Computational Biology*, vol. 5, e1000550.

Hase, T.; Niimura, Y. & Tanaka, H. (2010). Difference in gene duplicability may explain the difference in overall structure of protein-protein interaction networks among eukaryotes, *BMC Evolutionary Biology*, vol. 10, 358.

He, X. & Zhang, J. (2005). Rapid subfunctionalization accompanied by prolonged and substantial neofunctionalization in duplicate gene evolution, *Genetics*, vol. 169, pp. 1157–1164.

Ispolatov, I.; Krapivsky, PL.; Mazo, I. & Yuryev, A. (2005a). Cliques and duplication-divergence network growth, *New Journal of Physics*, vol. 7, 145.

Ispolatov, I.; Yuryev, A.; Mazo, I. & Maslov, S. (2005b). Binding properties and evolution of homodimers in protein-protein interaction networks, *Nucleic Acids Research*, vol. 33, pp. 3629–3635.

Ispolatov, I.; Krapivsky, PL. & Yuryev, R. (2005c). Duplication-divergence model of protein interaction network, *Physical Review E*, vol. 71, 061911.

Ito, T.; Chiba, T.; Ozawa, R.; Yoshida, M.; Hattori, M. & Sakaki, Y. (2001). A comprehensive two-hybrid analysis to explore the yeast protein interactome, *Proceeding of National Academy of Sciences USA*, vol. 98, pp. 4569–4574.

Jeong, H.; Mason, SP.; Barabasi, AL. & Oltvai, ZN. (2001). Lethality and centrality in protein interaction networks, *Nature*, vol. 411, pp. 41–42.

Jonsson, PF. & Bates, PA. (2006). Global topological features of cancer proteins in the human interactome, *Bioinformatics*, vol. 22, pp. 2291–2297.

Kim, J.; Krapivsky, PL.; Kahng, B. & Redner, S. (2002). Infinite-order precolation and giant fluctuations in a protein interaction network, *Physical Review E*, vol. 66, 055101.

Kola, I. & Landis, J. (2004). Can the phamaceutical industry reduce attrition rates?, *Nature Reviews Drug Discovery*, vol. 3, pp. 711–715.

Kuhn, M.; Campillos, M.; Gonzalez, P.; Jensen, LJ. & Bork, P. (2008). Large-scale prediction of drug-target relationships, *FEBS Letters*, vol. 582, pp. 1283–1290.

Kuhner, S.; Noort, VV.; Betts, MJ.; Leo-Macias, A.; Batisse, C.; Rode, M.; Yamada, T.; Maier, T.; Bader, S.; Beltran-Alvarez, P.; Castano-Diez, D.; Chen, WH.; Devos, D.; Guell, M.; Norambuena, T.; Racke, I.; Rybin, V.; Schmidt, A.; Yus, E.; Aebersold, R.; Herrmann, R.; Bottcher, B.; Frangakis, AS.; Russell, RB.; Serrano, L.; Bork, P. & Gavin, AC. (2009). Proteome organization in a genome-reduced bacterium, *Science*, vol. 326, pp. 1235–1240.

LaCount, DJ.; Vignali, M.; Chettier, R.; Phansalkar, A.; Bell, R.; Hesselberth, JR.; Schoenfeld, LW.; Ota, I.; Sahasrabudhe, S.; Kurschner, C.; Fields, S. & Hughes, RE. (2005). A protein interaction network of the malaria parasite Plasmodium falciparum, *Nature*, vol. 438, pp. 103–107.

Lavallee-Adam, M.; Cloutier, P.; Coulombe, B. & Blanchette, M. (2011) Modeling contaminants in AP-MS/MS experiments, *Journal of Proteome Research*, vol. 10, pp. 886–895.

Li, S.; Armstrong, CM.; Bertin, N.; Ge, H.; Milstein, S.; Boxem, M.; Vidalain, PO.; Han, JD.; Chesneau, A.; Hao, T.; Goldberg, DS.; Li, N.; Martinez, M.; Rual, JF.; Lamesch, P.; Xu, L.; Tewari, M.; Wong, SL.; Zhang, LV.; Berriz, GF.; Jacotot, L.; Vaglio, P.; Reboul, J.; Hirozane-Kishikawa, T.; Li, Q.; Gabel, HW.; Elewa, A.; Baumgartner, B.; Rose, DJ.; Yu, H.; Bosak, S.; Sequerra, R.; Fraser, A.; Mango, SE.; Saxton, WM.; Strome, S.; Heuvel, VDS.; Piano, F.; Vandenhaute, J.; Sardet, C.; Gerstein, M.; Doucette-Stamm, L.; Gunsalus, KC.; Harper, JW.; Cusick, ME.; Roth, FP.; Hill, DE. & Vidal, M. (2004). A map of the interactome network of the metazoan C. elegans, *Science*, vol. 303, pp. 540–543.

Liang, H. & Li, WH. (2007). Gene essentiality, gene duplicability and protein connectivity in human and mouse, *Trends in Genetics*, vol. 23, pp. 375–378.

Lynch, M. & Force, A. (2000). The probability of duplicate gene preservation by subfunctionalization, *Genetics*, vol. 154, pp. 459–473.

Maslov, S. & Sneppen, K. (2002). Specificity and stability in topology of protein networks, *Science*, vol. 296, pp. 910–913.

Middendorf, M.; Ziv, E. & Wiggins, CH. (2005). Inferring network mechanisms : the Drosophila melanogaster protein interaction network, *Proceedings of National Academy of Sciences USA*, vol. 102, pp. 3192–3197.

Moore, TJ.; Cohen, MR. & Furberg, CD. (2007). Serious adverse drug events reported to Food and Drug Administration, 1998-2005, *Archives of Internal Medicine*, vol. 167, pp. 1752–1759.

Newman, MEJ. (2002). Assortative mixing in networks, *Phyical Review Letters*, vol. 89, 208701.

Ohno, S. (1970). *Evolution by gene duplication*, Springer-Verlag, New-York, USA.

Pache, RA.; Zanozoni, A.; Naval, J.; Mas, JM. & Aloy, P. (2008). Towards a molecular characterisation of pathological pathways, *FEBS Letters*, vol. 582, pp. 1259–1265.

Pacifico, S.; Liu, G.; Guest, S.; Parrish, JR.; Fotouhi, F. & Finley, RL. (2006). A database and tool, IM Browser, for exploring and integrating emerging gene and protein interaction data for Drosophila, *BMC Bininformatics*, vol. 7, 195.

Parrish, JR.; Gulyas, KD. & Finley, RL. (2006) . Yeast two-hybrid contributions to interactome mapping, *Current Opinion in Biotechnology*, vol. 17, pp. 387–393.

Parrish, JR.; Yu, J.; Liu, G.; Hines, JA.; Chan, JE.; Mangiola, BA.; Zhang, H.; Pacifico, S.; Fotouhi, F.; DiRita, VJ.; Ideker, T.; Andrew, P. & Finley, RL. (2007). A proteome-wide protein interaction map for Campylobacter jejuni, *Genome Biology*, vol. 8, R130

Pastor-Satorras, R.; Vazquez, A. & Vespignani, A. (2001). Dynamical and correlation properties of the internet, *Physical Review Letters*, vol. 87, 258701.

Pastor-Satorras, R.; Smith, ED. & Sole, RV. (2003). Evolving protein interaction networks through gene duplication, *Journal of Theorerical Biology*, vol. 222, pp. 199–210.

Perlman, L.; Gottlieb, A.; Atias, N.; Ruppin, E. & Sharan, R. (2011). Combining drug and gene simiality measures for drug-target elucidation, *Journal of Computational Biology*, vol. 18, pp. 133–145.

Rambaldi, D.; Giorgi, FM.; Capuani, F.; Ciliberto, A. & Ciccarelli, FD. (2008). Low duplicability and network fragility of cancer genes, *Trends in Genetics*, vol. 24, pp. 427–430.

Rask-Andersen, M.; Almen, MS. & Schioth, HB. (2011). Trends in the exploitation of novel drug targets, *Nature Reviews Drug Discovery*, vol. 10, pp. 579–590.

Raval, A. (2003). Some asymptotic properties of duplication graphs, *Physical Review E*, vol. 68, 066119.

Reguly, T.; Breitkreutz, A.; Boucher, L.; Breitkreutz, BJ.; Hon, GC.; Myers, CL.; Parsons, A.; Friesen, H.; Oughtred, R.; Tong, A.; Stark, C.; Ho, Y.; Botstein, D.; Andrews, B.; Boone, C.; Troyanskya, OG.; Ideker, T.; Dolinski, K.; Batada, NN. & Tyers, M. (2006). Comprehensive curation and analysis of global interaction networks in Saccharomyces cerevisiae, *Journal of Biology*, vol. 5, 11.

Rual, JF.; Venkatesan, K.; Hao, T.; Hirozane-Kishikawa, T.; Dricot, A.; Li, N.; Berriz, GF.; Gibbons, FD.; Dreze, M.; Ayivi-Guedehoussou, N.; Klitgord, N.; Simon, C.; Boxem,

M.; Milstein, S.; Rosenberg, J.; Goldberg, DS.; Zhang, LV.; Wong, SL.; Franklin, G.; Li, S.; Albala, JS.; Lim, J.; Fraughton, C.; Llamosas, E.; Cevik, S.; Bex, C.; Lamesch, P.; Sikorski, RS.; Vandenhaute, J.; Zoghbi, HY.; Smolyar, A.; Bosak, S.; Sequerra, R.; Doucette-Stamm, L.; Cusick, ME.; Hill, DE.; Roth, FP. & Vidal, M. (2005). Towards a proteome-scale map of the human protein-protein interaction network, *Nature*, vol. 437, pp. 1173-1178.

Shimoda, Y.; Shinpo, S.; Kohara, M.; Nakamura, Y.; Tabata, S. & Sato, S. (2008). A large scale analysis of protein-protein interactions in Nitrogen-fixing Bacterium Mesorhizobium loti, *DNA Research*, vol. 15, pp. 13-23.

Simonis, N.; Rual, JF.; Carvunis, AR.; Tasan, M.; Lemmens, I.; Hirozane-Kishikawa, T.; Hao, T.; Sahalie, JM.; Venkatesan, K.; Gebreab, F.; Cevik, S.; Klitgord, N.; Fan, C.; Braun, P.; Li, N.; Ayivi-Guedehoussou, N.; Dann, E.; Bertin, N.; Szeto, D.; Dricot, A.; Yildirim, MA.; Lin, C.; Smet, AS.; Kao, HL.; Simon, S.; Smolyar, A.; Ahn, JS.; Tewari, M.; Boxem, M.; Milstein, S.; Yu, H.; Dreze, M.; Vandenhaute, J.; Gunsalus, KC.; Cusick, ME.; Hill, DE.; Tavernier, J.; Roth, FP. & Vidal, M. (2009). Empirically controlled mapping of the Caenorhabditis elegans protein-protein interactome network, *Nature Methods*, vol. 6, pp. 47-54.

Sole, RV.; Pastor-Satorras, R.; Smith, ED. & Kepler, T. (2002). A model of large-scale proteome evolution, *Advances in Complex Systems*, vol. 5, pp. 43-54.

Stelzl, U.; Worm, U.; Lalowski, M.; Haenig, C.; Brembeck, FH.; Goehler, H.; Stroedicke, M.; Zenkner, M.; Schoenherr, A.; Koeppen, S.; Timm, J.; Mintzlaff, S.; Abraham, C.; Bock, N.; Kietzmann, S.; Goedde, A.; Toksöz, E.; Droege, A.; Krobitsch, S.; Korn, B.; Birchmeier, W.; Lehrach, H. & Wanker, EE. (2005). A human protein-protein interaction network: a resource for annotating proteome, *Cell*, vol. 122, pp. 957-968.

Titz, B.; Rajagopala, SV.; Goll, J.; Hauser, R.; McKevitt, MT.; Palzkill, T. & Uetz, P. (2008). The binary protein interactome of Treponema pallidum — the syphilis spirochete, *PLoS ONE*, vol. 3, e2292.

Uetz, P.; Giot, L.; Cagney, G.; Mansfield, TA.; Judson, RS.; Knight, JR.; Lockshon, D.; Narayan, V.; Srinivasan, M.; Pochart, P.; Qureshi-Emili, A.; Li, Y.; Godwin, B.; Conover, D.; Kalbfleisch, T.; Vijayadamodar, G.; Yang, M.; Johnston, M.; Fields, S. & Rothberg, JM. (2000). A comprehensive analysis of protein-protein interactions in Saccharomyces cerevisiae, *Nature*, vol. 403, pp. 623-627.

Vazquez, A.; Flammini, A.; Maritan, A. & Vespignani, A. (2003). Modeling of protein interaction networks, *Complexus*, vol. 1, pp. 38-44.

Vazquez, A. (2003). Growing networks with local rules : preferential attachment, clustering hierarchy, and degree correlations, *Physical Review E*, vol. 67, 056104.

Venkatesan, K.; Rual, JF.; Vazquez, A.; Stelzl, U.; Lemmens, I.; Hirozane-Kishikawa, T.; Hao, T.; Zenkner, M.; Xin, X.; Goh, KI.; Yildirim, MA.; Simonis, N.; Heinzmann, K.; Gebreab, F.; Sahalie, JM.; Cevik, S.; Simon, C.; Smet, AS.; Dann, E.; Smolyar, A.; Vinayagam, A.; Yu, H.; Szeto, D.; Borick, H.; Dricot, A.; Klitgord, N.; Murray, RR.; Lin, C.; Lalowski, M.; Timm, J.; Rau, K.; Boone, C.; Braun, P.; Cusick, ME.; Roth, FP.; Hill, DE.; Tavernier, J.; Wanker, EE.; Barabasi, AL. & Vidal, M. (2009). An empirical framework for binary interactome mapping, *Nature Methods*, vol. 6, pp. 83-90.

Vidal, M.; Cusick, ME. & Barabasi, AL. (2011). Interactome networks and human disease, *Cell*, vol. 144, pp. 986-998.

Wachi, S.; Yoneda, K. & Wu, R. (2005). Interactome-transcriptome analysis reveals the high centrality of genes differentially expressed in lung cancer tissues, *Bioinformatics*, vol. 21, pp. 4205–4208.

Wang, J.; Li, ZX.; Qui, CX.; Wang, D. & Cui, QH. (2011). The relationship between rational drug design and drug side effects, *Briefings in Bioinformatics*, (in press).

Wagner, A. (2002). Asymmetric functional divergence of duplicate genes in yeast, *Molecular Biology and Evolution*, vol. 19, pp. 1760–1768.

Wagner, A. (2003). How the global structure of protein interaction networks evolves, *Proceedings of the Royal Society B: Biological Sciences*, vol. 270, pp. 457–466.

Watts, DJ. & Strogatz, SH. (1998). Collective dynamics of 'small-world' networks, *Nature*, vol. 393, pp. 440–442.

Xu, J. & Li, Y. (2006). Discovering disease-genes by topological features in human protein-protein interaction network., *Bioinformatics*, vol. 22, pp. 2800–2805.

Yamanishi, Y.; Kotera, M.; Kanehisa, M. & Goto, S. (2010). Drug-target interaction prediction from chemical, genomic and pharmacological data in an integrated framework, *Bioinformatics*, vol. 26, pp. i246–254.

Yao, L. & Rzhetsky, A. (2008). Quantitative systems-level determinants of human genes targeted by successful drugs, *Genome Research*, vol. 18, pp. 216–213.

Yildirim, MA.; Goh, KI.; Cusick, ME.; Barabasi, AL. & Vidal, M. (2007). Drug-target network, *Nature Biotechnology*, vol. 25, pp. 1119–1126.

Yu, H.; Braun, P.; Yildirim, MA.; Lemmens, I.; Venkatesan, K.; Sahalie, J.; Hirozane-Kishikawa, T.; Gebreab, F.; Li, N.; Simonis, N.; Hao, T.; Rual, JF.; Dricot, A.; Vazquez, A.; Murray, RR.; Simon, C.; Tardivo, L.; Tam, S.; Svrzikapa, N.; Fan, C.; de Smet, AS.; Motyl, A.; Hudson, ME.; Park, J.; Xin, X.; Cusick, ME.; Moore, T.; Boone, C.; Snyder, M.; Roth, FP.; Barabási, AL.; Tavernier, J.; Hill, DE. & Vidal, M. (2008). High-quality binary protein interaction map of the yeast interactome network, *Science*, vol. 322, pp. 104–110.

Zanzoni, A.; Soler-Lopez, M. & Aloy, P. (2009). A network medicine approach to human disease, *FEBS Letters*, vol. 583, pp. 1759–1765.

Zhao, S. & Li, S. (2010). Network-based relating pharmacological and genomic spaces for drug target identification, *PLoS ONE*, vol. 5, e11764.

6

Computational Tools and Databases for the Study and Characterization of Protein Interactions

Jose Ramon Blas[1], Joan Segura[2] and Narcis Fernandez-Fuentes[2,3]
[1]*Universidad de Castilla-La Mancha*
[2]*University of Leeds*
[3]*Aberystwyth University*
[1]*Spain*
[2,3]*United Kingdom*

1. Introduction

One of the most pressing challenges in the post genomic era is the characterization and charting of protein-protein interactions (PPIs) in living organisms, as these are essential in the shaping of normal and pathological behaviours in cells. It is for this reason that unravelling the nature of PPIs has been the pursuit of many experimental techniques, ranging from high-throughput to high-detail approaches (Shoemaker and Panchenko 2007), as well as a wide spectrum of computational prediction methods. Current estimations of human interactome size range from 100,000 to more than 600,000 interactions (Bork et al. 2004; Stelzl and Wanker 2006; Stumpf et al. 2008; Venkatesan et al. 2009). Experimental strategies have reached their best at describing around 50,000 interactions by collating a large number of small and very focused experiments with high-throughput ones, such as massive yeast two-hybrid (Rual et al. 2005; Stelzl et al. 2005), or mass spectrometry coupled to affinity purification experiments (Ewing et al. 2007; Hubner et al. 2010). The smallest gap between experimentally validated and theoretically predicted PPIs amounts to around 50% of total interactions, being probably much higher. When it comes to to studying PPIs in other species on which, even having been sequenced, experimental data is even more scarce, the need for PPI-map completeness is even more notorious. Computational prediction and characterization of PPIs, with its drawbacks, successes and challenges, constitutes a valuable aid in the way to a complete description of interactomes, hence being a promising research field that has enriched our image of living cells for some time now.

Computational tools can provide useful information at different levels of resolution and this chapter seeks to present an up-to-date and comprehensive review of these. The first part of the chapter presents the theoretical basis of computational tools designed to predict PPIs. The main aim of these tools is to predict whether two proteins A and B can interact, either directly or indirectly (functional associations), but without dwelling on the molecular details of the interaction, i.e. which proteins interact. These predictions are useful as complement to large-scale experimental analyses, either to confirm observed interactions or discard false positives, and also to uncover novel interactions. The second part of the chapter is devoted to the computational methods developed to predict protein interfaces. At this level, predictions identify specific regions and residues of the protein that are likely to mediate PPIs. Thus, these methodologies uncover a higher level of detail, i.e. *how* proteins interact, and have a number of applications in experimental work such as guiding the mapping of protein interfaces by mutagenesis or structural modelling of protein complexes. A special emphasis will be given to a novel and highly accurate tool: VORFFIP(Segura et al. 2011). The concluding part of the chapter describes computational tools developed to predict the important regions or *hot spots* in protein interfaces. Recent successes in the quest for finding new therapeutic agents to modulate PPIs have been aided by the realization, following the pioneering work by Clackson and Wells(Clackson and Wells 1995), that the binding energy of many PPIs can be ascribed to a small and complementary set of interfacial residues: a *hot spot* of binding energy. Thus, identifying these critical residues by computational means has clear applications in drug discovery and in some aspects of protein design. PCRPi(Assi et al. 2010), a novel and highly precise tool will be discussed.

2. Prediction of protein-protein interactions

With the aim of detailing a complete protein interaction map that agglutinates the rising amount of genomic data, high-throughput experimental techniques have walked in parallel with computational approaches. There are six basic computational approaches to predict PPIs depending on the nature of the information used for the prediction. These include PPIs inferred from: (i) genomic context including phylogenetic profiles, gene neighbouring analyses and gene fusion events; (ii) co-evolution events; (iii) protein domain co-occurrence (or signatures) between pair of proteins; (iv) text mining; (v) transference of annotation between species: protein-protein interologs; and (vi) structural annotation including homology-based or *ab initio*. Figure 1 depicts an overall diagrammatic description of these basic approaches and tables 1 and 2 compile a number of on line databases and computational tools respectively.

2.1 Genomic context methods

Biological processes subjected to evolutionary pressure tend to cluster together all interrelated molecular actors in single units to simplify control mechanisms and thus avoid the lost of any essential component. This principle, which operates from maintaining bacterial operon systems to more sophisticated co-regulation strategies found in eukaryotes, is on the basis of genomic context based methods for the detection of functional PPIs.

The first group of genome context methods are based in the comparison of phylogenetic profiles. A phylogenetic profile is the presence or absence of a given gene across N species that can be expressed as an N-dimensional array of ones and zeroes. Originally, functional relationship was assumed if having similar phylogenetic profiles(Pellegrini et al. 1999); however, positive results were limited to very strong interactions and many relations between analogous proteins were missing. Further improvements were made by discarding overlaps given by chance(Wu et al. 2003), using protein domains instead of full length proteins(Pagel et al. 2004), through a concurrent search of multiple independent phylogenetic events of gain/loss of pairs of genes to discard spurious correlated patterns(Barker and Pagel 2005), or the use of enhanced representation of phylogenetic trees(Ta et al. 2011). The second group of genome context methods is based on gene closeness among different genomes, considering closeness as a sign of functional relatedness. After some initial successes(Koonin et al. 2001; Evguenieva-Hackenberg et al. 2003) and despite some improvements such as allowing for changes in gene order and orientation(Szklarczyk et al. 2011), large-scale predictions should be considered cautiously. Finally, gene fusion approaches are a group of computational tools based on the evidence that some interacting proteins have orthologous where both proteins appear fused in a single protein. Thus, it has been observed that many of these pairs, fused in single proteins in other organisms, correspond to binding partners or at least functionally related proteins(Marcotte et al. 1999; Yanai et al. 2001). Rosetta method(Marcotte et al. 1999) and other implementations(Enright et al. 1999) exploit gene fusion events as predictors of PPIs.

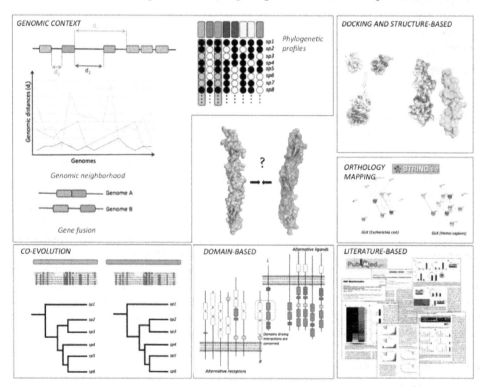

Fig. 1. Graphical description of the main strategies of PPIs prediction described in sections 2.1 to 2.6.

2.2 Co-evolution methods

Two proteins that share a functional relationship, either through direct interaction or functional association, may present evidences of co-evolution. Since the seminal work of Altschuh identifying correlated amino acid changes in Tobacco mosaic virus(Altschuh et al. 1987), studies recognizing co-evolution as an indicative, albeit subtle, signal of PPIs have being reported in the literature (Travers and Fares 2007; Chao et al. 2008; Presser et al. 2008). Co-evolutionary information may be divided into three groups: the simultaneous loss or gain of orthologous genes(Marcotte et al. 1999), correlated changes affecting both interacting partners at whole sequence level (explored by *mirrortree*-based approaches)(Goh and Cohen 2002; Hakes et al. 2007; Juan et al. 2008) or single amino acids changes(Mintseris and Weng 2005; Madaoui and Guerois 2008).

In the case of mirrortree-based methods (e.g.(Ochoa and Pazos 2010)), the likelihood of interaction is measured as a correlation value between the phylogenetic trees of two families of proteins. Although these approaches have been successfully applied in PPIs prediction(Labedan et al. 2004; Dou et al. 2006; McPartland et al. 2007; Juan et al. 2008), it is still a major problem distinguishing between co-evolution arising from a direct PPI, what has been termed as *co-adaptation* (Pazos and Valencia 2008), from non-specific changes and thus not necessary driven by a functional relatedness(Lovell and Robertson 2010). Recent advances in this area include MatrixMatchMaker algorithm(Tillier and Charlebois 2009) and a faster implementation suitable for large-scale analyses(Rodionov et al. 2011).

The detection of site-specific co-evolution events reflecting PPIs, despite being more intuitive and informative, is even more challenging given the complexity of the mixed evolutionary-structural scenario involved. A single point mutation might ease or complicate each imaginable path of mutation at any other position in the complex, regardless of its distance from the interface(Lovell and Robertson 2010). In fact, co-evolution events have been detected affecting sites that are distant structurally (Gobel et al. 1994; Clarke 1995; Gloor et al. 2005; Fares and McNally 2006). On the other hand, the probability of correlated amino acid changes is closely related to the chemical nature of changes. In this sense, volume variations seem to strongly affect fitness, and so they are frequently balanced by evolution machinery (up to almost 50% of the cases)(Williams and Lovell 2009). Moreover, interface residues in obligate complexes evolve at a slower rate than those in transient interactions(Mintseris and Weng 2005).Taken together, all these particulars illustrate the challenges encountered when looking for site-specific co-evolution events related to PPIs. Recent developments have looked at improving the discrimination between direct and indirect correlations(Burger and van Nimwegen 2010), or including amino acid background distribution information and the mutual information of residues physicochemical properties(Gao et al. 2011). However, new, more discriminative, approaches are required to better understand co-evolution at residue-centred level.

2.3 Domain-based methods

There are strong evidences supporting the idea that the range of different PPIs can be accounted for by considering a more reduced set of specific domain-domain interactions, domain signatures, that are even conserved across different species(Finn et al. 2006; Itzhaki et al. 2006; Stein et al. 2011). Thus, the basis of domain-based methods is presence/absence of given domain signatures between pairs of proteins that can be used to infer interaction. An early method exploiting domain signatures was an association method where domain interactions were assumed if the frequency of association was higher than the expected frequency(Kim et al. 2002). Further improvements have been devised to improve predictions including the domain pair exclusion analysis, which implemented a new scoring scheme(Riley et al. 2005), the use of Random Forest ensemble classifiers to deal with the pairing of multi-domain proteins(Chen and Liu 2005) or the use of Gene Ontology(Lee et al. 2006) or co-evolution data(Jothi et al. 2006).

2.4 Literature-based data mining methods

Numerous research efforts have been focused on automatically extracting and analysing information from the scientific literature in order to infer putative PPIs(Blaschke et al. 2001; Fundel et al. 2007; Airola et al. 2008). These include, the search for the co-occurrence of terms(Blaschke et al. 2001) or the presence of similar Gene Ontology terms(Pesquita et al. 2009) or kernel-based methods including subsequence kernels, tree kernels, shortest path kernels and graph kernels(Tikk et al. 2010). The most recent approaches use multiple kernels to maximize the information extracted from scientific papers(Kim et al. 2008; Miwa et al. 2009), the combination of multiple kernels and machine learning algorithms to improve the scoring(Yang et al. 2011), or the more recent neighbourhood hash graph kernels that are substantially faster than previous text-mining approaches(Zhang et al. 2011).

Name	URL	Reference
STRING	http://string-db.org	(Szklarczyk et al. 2011)
BioGRID	http://thebiogrid.org/	(Stark et al. 2011)
IntAct	http://www.ebi.ac.uk/intact/	(Aranda et al. 2010)
HPRD	http://www.hprd.org/	(Prasad et al. 2009)
HitPredict	http://hintdb.hgc.jp/htp/	(Patil et al. 2011)
DIP	http://dip.doe-mbi.ucla.edu/dip	(Salwinski et al. 2004)
MINT	http://mint.bio.uniroma2.it/mint/	(Chatr-aryamontri et al. 2007)
TAIR	www.arabidopsis.org/portals/proteome/	(Swarbreck et al. 2008)
iPFAM	http://ipfam.sanger.ac.uk/	(Finn et al. 2005)
3DID	http://3did.irbbarcelona.org/	(Stein et al. 2011)
DIMA 3.0	http://webclu.bio.wzw.tum.de/dima/	(Luo et al. 2011)
DOMINE	http://domine.utdallas.edu/cgi-bin/Domine	(Yellaboina et al. 2011)
GWIDD	http://gwidd.bioinformatics.ku.edu	(Kundrotas et al. 2010)
IsoBase	http://isobase.csail.mit.edu/	(Park et al. 2011)
I2D	http://ophid.utoronto.ca/ophidv2.201	(Brown and Jurisica 2007)
DroID	http://www.droidb.org	(Murali et al. 2011)
HCPIN	http://nesg.org:9090/HCPIN	(Huang et al. 2008)
HIV1,HPID	http://www.ncbi.nlm.nih.gov/RefSeq/HIVInteractions	(Fu et al. 2009)
MPIDB	http://www.jcvi.org/mpidb/about.php	(Goll et al. 2008)

Table 1. List of major databases compiling experimentally determined or computationally predicted PPIs.

2.5 Orthology mapping (Interologs) methods

The basis of these methods is the transference of annotated interactions between organisms; hence the term *interologs* to refer to predicted homologous interactions(Walhout et al. 2000; Shoemaker and Panchenko 2007; Lewis et al. 2010). Interolog annotations have been successfully applied to transfer experimentally known interactions in yeast to predicted ones in worm(Matthews et al. 2001) and between mouse and human(Huang et al. 2007). Although some improvement has been devised such as scoring schemes that depend on the sources of experimental data(Jonsson and Bates 2006), the applicability of orthology mapping is limited. Firstly, accurate predictions require high sequence similarities between interologs (~70%)(Mika and Rost 2006) thus limiting its range of applicability. Secondly, even at high sequence identity level, in some cases small variations in protein sequence at the interface have been shown to dramatically change PPI specificity, thus redefining complex protein networks and leading to important phenotypic differences(Panni et al. 2002; Kiemer and Cesareni 2007).

2.6 Structure-based methods

A final category of computational methods includes those based in structural information. The structure of a protein complex formed by two or more proteins can be modelled using the structure of a known protein complex as template either by homology modelling or threading(Lu et al. 2002; Aloy et al. 2004; Hue et al. 2010). Even in the absence of a suitable template, the structure of the complex can be modelled by using protein docking(Wass et al. 2011) and selecting the protein complex based on predicted binding energy, i.e. *ab initio* modeling. Despite being a promising strategy, and without considering the high computational cost, the correlation between predicted and experimentally measured binding affinities, such as K_d, is very low thus greatly impairing its predictive power (Kastritis and Bonvin 2010; Stein et al. 2011). Other strategies combine structural data, docking and evolutionary conservation(Tuncbag et al. 2011).

Name	Methodology	URL	Reference
MirrorTree	Co-evolution	http://csbg.cnb.csic.es/mtserver/	(Ochoa and Pazos 2010)
MatrixMatchMaker	Co-evolution	http://www.uhnresearch.ca/labs/tillier/	(Tillier and Charlebois 2009)
iHOP	Text mining	http://www.ihop-net.org/	(Hoffmann and Valencia 2004)
PathBLAST	Orthology	http://www.pathblast.org/	(Kelley and Ideker 2005)
InterPreTS	Structure-based	http://www.russelllab.org/	(Aloy and Russell 2003)
IBIS	Structure-based	http://www.ncbi.nlm.nih.gov/Structure/ibis/ibis.cgi	(Shoemaker et al. 2010)

Table 2. List of on line resources for the prediction of PPIs.

3. Prediction of protein binding sites

As indicated by its name, binding site prediction methods seek to define the regions in proteins that are more likely to mediate PPIs. The level of resolution is therefore higher and the starting point is either the sequences or structures of proteins that are known to interact (i.e. experimental evidence) but for which no structural details of the interaction are known.

3.1 Distinctiveness of interface residues

Large-scale analyses of the structures of protein complexes have shown that residues located in interfaces present a number of differential physicochemical and structural qualities. In general, hydrophobic residues are overrepresented in the interfaces of permanent complexes(Lo Conte et al. 1999; Glaser et al. 2001) and charge residues, Arg in particular, are also commonly found in interfaces and often define the lifetime of complexes(Zhou and Shan 2001). A higher accessibility to the solvent than exposed residues not located in interfaces is also a differential trait of interface residues(Chen and Zhou 2005), being the most effective feature to predict interfaces in homodimeric complexes(Jones and Thornton 1997). On the other hand and in agreement with earlier observations that found interface residues have lower crystallographic B-factors(Neuvirth et al. 2004), the side chains of interface residues are less likely to sample alternative rotamers, i.e. more rigid, to decrease the entropic cost upon complex formation(Cole and Warwicker 2002; Fleishman et al. 2011). Sequence conservation has also proved to be a predictor(Lichtarge et al. 1996; Wang et al. 2006), although it remains a contentious issue as some works have shown that interfaces are not more conserved than the rest of the protein(Grishin and Phillips 1994; Caffrey et al. 2004). Finally, it has been shown that interfaces are richer in β-strands and long loops while α-helical conformations are disfavoured(Neuvirth et al. 2004).

3.2 Prediction methods

Prediction methods rely on sequence and/or structural information that is unique to interface residues (see before). Hence, prediction methods can be divided into two groups: sequence-based methods, which rely only on the primary sequence of the protein and structure-based methods that require the three-dimensional structure of the protein. Table 3 compiles a list of on line computational tools to predict protein binding sites.

3.2.1 Structure-based prediction methods

One of the first structure-based prediction methods, later updated(Murakami and Jones 2006), was based on surface patch analysis(Jones and Thornton 1997). Surface patches were defined by grouping neighbouring exposed residues that were subsequently ranked using a scoring function that included the solvation potential, interface propensity, hydrophobicity, protrusion and accessible surface area of each of the residues within the patch. A probabilistic approach, ProMate, also based on patch analysis, was developed for heteromeric transient protein complexes by combining secondary structure content, hydrophobicity and crystallographic B-factors information(Neuvirth et al. 2004). The combination of ProMate's predictions and a parametric scoring function based of sequence conservation and structural features resulted in an improvement of the accuracy of the predictions(de Vries et al. 2006). Other implementations of prediction methods include an empirical scoring function composed of side chain energy score, residue conservation and interface propensity(Liang et al. 2006), the search of structural interaction templates extracted from protein complexes(Chang et al. 2006) and a clustering algorithm that identifies residues with a high propensity of being located in interfaces(Negi et al. 2007).

In order to combine and integrate heterogenous data, i.e. sources of information of a different nature (e.g. hydrophobicity indexes and solvent accessibility surface) into a common and coherent scoring framework, a number of machine learning methods have been proposed including Neural Networks (NN)(Fariselli et al. 2002; Chen and Zhou 2005; Porollo and Meller 2007), Support Vector Machines (SVM)(Bradford and Westhead 2005), Random Forests (RF)(Sikic et al. 2009; Segura et al. 2011) and Bayesian Networks (BN)(Bradford et al. 2006; Ashkenazy et al. 2010). Thus, the commonality of these approaches is the use of a machine-learning algorithm (NN, SVM, RF or BN) to combine a set of sequence- and structural-based measures into an unified score or probability. The nature of the combined features used by the prediction methods includes: evolutionary conservation and surface disposition(Fariselli et al. 2002); sequence conservation, electrostatic potentials, SASA, hydrophobicity, protusion and interface propensity(Bradford and Westhead 2005; Bradford et al. 2006); properties taken from the AAIndex database(Kawashima et al. 2008) (e.g. expected number of contacts within 14 Å sphere), multiple sequence alignment-derived features (e.g. amino acid frequency), and structural features(Porollo and Meller 2007); structure-based, energy terms, sequence conservation and crystallographic B-factors(Segura et al. 2011); structural features, sequence and secondary structure(Sikic et al. 2009); or more complex approaches that combine several prediction methods in the form of a meta-prediction(Qin and Zhou 2007; Ashkenazy et al. 2010).

3.2.2 Sequence-based prediction methods

Even if the structure of the protein is not available, there are still a number of prediction methods that are based solely on sequence information. Early examples of approaches in this category include a NN (Ofran and Rost 2003) that uses local sequence information, which was subsequently improved by including a post-neural network filtering step(Ofran and Rost 2007). Other approaches include SVMs that combine sequence profiles and other sequence-based information such as spatially neighbouring residues(Koike and Takagi 2004; Res et al. 2005; Chen and Li 2010), a RF that integrates physicochemical properties of residues, evolutionary conservation and amino acid distances(Chen and Jeong 2009), and a naive Bayesian classifier trained to integrate position-specific scoring matrix and predicted accessibility(Murakami and Mizuguchi 2010). Finally, other sequence-based methods have been developed to improve prediction by tacking issues such as the problem of unbalanced data in protein sets(Yu et al. 2010), i.e. the interface accounts for a small proportion of the exposed residues so the number of negative cases (non-interface residues) is much larger than the number of positive cases (interface residues) or improving the sampling(Engelen et al. 2009) in evolutionary trace-based(Lichtarge et al. 1996) methodologies.

3.3 VORFFIP, a holistic approach to predict protein binding sites in protein structures

VORFFIP is a novel, structure-based, method that integrates a wide range of residue-based features and environment information using a 2-step Random Forest ensemble classifier(Segura et al. 2011). Residue-based features include structural-based, energy terms, evolutionary conservation and crystallographic B-factors information. VORFFIP implements a novel definition of local environment by means of Voronoi Diagrams (see next and Fig. 2) that complements residue-based information improving the accuracy of predictions.

Residue-based information characterizes individual residues. Structure-based features account from 16 different features and define the local geometry of the protein at residue level. Structural features include, among others, the absolute and relative accessibility surface area, the protrusion index that is a measure of the local concavity/convexity and a deepness index(Vlahovicek et al. 2005). The energetic state of exposed residue is characterized by 10 energy terms including electrostatic potential, solvent exposure energy, entropy and hydrogen bond energy among others(Guerois et al. 2002). The sequence conservation of residues consist of the regional conservation score that defines the conservation for each residue and its neighbourhood in the 3D space(Landgraf et al. 2001) and a sequence positional score calculated from multiple sequence alignment profiles(Pei and Grishin 2001). Finally, crystallographic B-factors, which are a measure of thermal motion, are converted to Z-score as described previoulsy(Yuan et al. 2003).

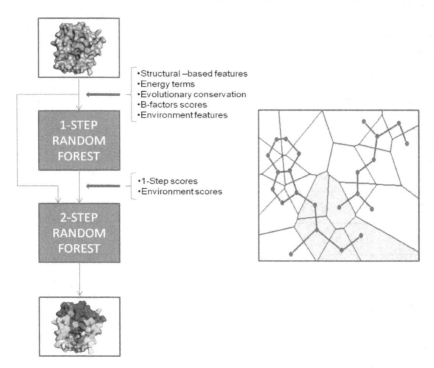

Fig. 2. Overview of prediction process in VORFFIP and a Voronoi Diagram of a interacting pair. The left side of the figure illustrates the 2-step prediction approach in VORFFIP. The right side of the figure shows the Voronoi Diagram of two neighbouring residues; heavy atoms are represented by red dots and coloured cells illustrate interaction between atoms of neighbouring residues.

Environment-based information accounts the local structural environment of residues. Interfaces tend to form contiguous patches on the surface and thus, the environment of a residue can provide valuable information for predictions. Several methods have been used to account for the local environment of residues including sliding window (e.g.(Ofran and Rost 2003)) and Euclidian distances (e.g. (Porollo and Meller 2007)). VORFFIP however uses a novel definition of environment by means of Voronoi Diagrams (VD). VD is computed using the heavy atoms coordinates as seeds and as a result the 3D space is partitioned into polyhedral cells where each single cell contains one of the atoms (Barber et al. 1996). Atoms sharing a common facet in the VD are said to be in contact or neighbours, i.e. part of the local environment. Figure 2 shows a 2D representation of a VD diagram depicting the interaction between atoms of two neighbouring residues. The number of contacts between

Name	Input	Method	URL	Reference
VORFFIP	Structure	RF	http://www.bioinsilico.org/VORFFIP	(Segura et al. 2011)
ProMate	Structure	Scoring function	http://bioinfo.weizmann.ac.il/promate	(Neuvirth et al. 2004)
ISIS	Sequence	NN	http://rostlab.org/cms/resources/web-services/	(Ofran and Rost 2007)
WHISCY	Structure	Scoring function	http://nmr.chem.uu.nl/Software/whiscy	(de Vries et al. 2006)
PPI-pred	Structure	SVM	http://www.bioinformatics.leeds.ac.uk/ppi_pred	(Bradford and Westhead 2005)
SPPIDER	Structure	NN	http://sppider.cchmc.org	(Porollo and Meller 2007)
PINUP	Structure	Scoring function	http://sparks.informatics.iupui.edu	(Liang et al. 2006)
meta-PPISP	Structure	Meta-server	http://pipe.scs.fsu.edu/meta-ppisp.html	(Qin and Zhou 2007)
Protemot	Structure	Scoring function	http://bioinfo.mc.ntu.edu.tw/protemot	(Chang et al. 2006)
InterProSurf	Structure	Scoring function	http://curie.utmb.edu	(Negi et al. 2007)
cons-PPISP	Structure	NN	http://pipe.scs.fsu.edu/ppisp.html	(Chen and Zhou 2005)
PSIVER	Sequence	BN	http://tardis.nibio.go.jp/PSIVER/	(Murakami and Mizuguchi 2010)
SHARP	Structure	Scoring function	http://www.bioinformatics.sussex.ac.uk/SHARP2	(Murakami and Jones 2006)

Table 3. List of online resources for protein binding site prediction.

neighbouring residues is used to derive weights that will be then used to normalize residue-based features among residues within the local environment. The advantage of using VD over other definition of local environment is that there are no requirements with regards cut-off to define the local environment (e.g. a distance cut-off) and that a weighting system can be easily implemented based on the number of interactions (i.e. neighbouring residues) in the VD. When the performance of VORFFIP was assessed in term of type of methods used to define local environment, VD were superior to Euclidean distances and sliding window approaches(Segura et al. 2011).

The final stage of the method is the integration of residue- and environment-based features using a machine learning approach: a 2-steps RF ensemble classifier (Fig. 2), which is also a novel feature as most machine learning methodologies use a single step classifier. In the first-step RF, residue and residue-environment features are calculated and used as input variables. The scores yielded by the first-step RF are then decomposed into a number of new input variables including VD-derived environment scores. Residue and environment scores together with the previously calculated features form the new set of input variables to the second-step RF that will output the final scores. The logic behind using a second-step RF relates to the observation that residues belonging to the same interface tend to form contiguous patches on the surface, i.e. high scoring residues are expected to be neighbouring mainly high scoring residues unless located at the boundaries of the interface. Thus, the second-step RF harmonizes outliers and generates more homogenous scores for interface residues resulting in better predictions as shown by the competitive results obtained(Segura et al. 2011) when comparing to other methods(de Vries et al. 2006; Porollo and Meller 2007; Sikic et al. 2009).

4. Prediction and charting of hot spots in protein interfaces

The final part of the chapter describes the current state in computational prediction of hot spots in protein interfaces. The goal of these methods is the prediction of the region of a given interface that contributes the most to the binding energy of the complex, i.e. the hot spot of the interaction. These methods are a good complement to highly intensive and costing experimental techniques, in particular in large-scale analyses, and have clear applications in drug discovery and protein engineering.

4.1 Distinctiveness of hot spot residues

As in the case of interface residues, hot spot residues present a number of structural and physicochemical properties unique to them and these are exploited by the prediction methods. The first is the type of residues that are commonly found in hot spots: while the proportion of Trp, Arg and Tyr is higher, Leu, Ser and Val are disfavoured(Bogan and Thorn 1998). Likewise, Asn and Asp are more commonly found in hot spots than chemically comparable (but bulkier) Gln and Glu(Bogan and Thorn 1998). Hot spot residues are optimally packed, structurally conserved and usually located in the central part of the interface(Keskin et al. 2005; Yogurtcu et al. 2008). One more characteristic of hot spot residues is that they are often located in complemented pockets, i.e. hot spot residues in one protein interact with hot spot residues of cognate protein(s)(Li et al. 2004). Finally, hot spot residues usually have a higher evolutionary conservation than the rest of the residues in the interface(Guharoy and Chakrabarti 2005).

4.2 Prediction algorithms

A number of computational methods have been developed for the prediction of hot spots in protein interfaces. An important part of these is represented by energy-based methods that predict changes in binding energy upon mutations, i.e. *in silico* alanine scanning. These methodologies range from scoring function derived from simple physical models(Guerois et al. 2002; Kortemme and Baker 2002; Kruger and Gohlke 2010) to more complex, time consuming atomistic simulations to model effect of mutations in the binding energy(Almlof et al. 2006; Lafont et al. 2007; Moreira et al. 2007; Benedix et al. 2009; Diller et al. 2010). Other methods exploit individual features (or combination of them) that are characteristic to hot spots such as solvent accessibility(Landon et al. 2007; Tuncbag et al. 2009; Xia et al. 2010; Li et al. 2011), atomic contacts(Li et al. 2006), structural conservation(Li et al. 2004), restricted mobility(Yogurtcu et al. 2008), relative location of residues in the interface(Keskin et al. 2005), sequence conservation(Hu et al. 2000; Ma and Nussinov 2007) and pattern mining(Hsu et al. 2007). Other examples include a number of machine learning approaches (Darnell et al. 2007; Ofran and Rost 2007; Cho et al. 2009; Lise et al. 2009; Assi et al. 2010) such as PCRPi (see next) that integrate a range of structural- and sequence-based information and a docking-based approach(Grosdidier and Fernandez-Recio 2008).

4.3 PCRPi: *Presaging Critical Residues in Protein interfaces*, a novel and highly accurate prediction algorithm

While the attributes described in section 4.1 have predictive power, it has been found that individually cannot unambiguously define hot spot residues(DeLano 2002). To overcome this limitation, PCRPi(Assi et al. 2010), a novel computation tool for the prediction of hot spots residues, integrates seven different variables that account for structural, evolutionary conservation and predicted binding energy (Fig.3).

The structural information of interface residues is described by two different variables: the interaction engagement (IE) and the topographical (TOP) indexes. The IE index gauges for the number of inter-chain atomic interactions of the given residue normalized by total number of atoms that can potentially interact. An IE index of 1.0 would indicate that all atoms are actively engaged in atomic interactions with groups of cognate protein(s). The TOP index describes the structural environment of residues and is ratio between the number of neighbouring residues of cognate proteins and the average number neighbouring residues. Neighbouring residues are any residues of cognate protein(s) whose carbon alpha is enclosed in a sphere of 10 Angstroms of radius centered on the carbon alpha of the residue of interest. Thus, TOP index quantifies whether residues are intimately interacting with cognate proteins or are located in a more flat or unprotected region.

The second group of variables used by PCRPi relates to evolutionary conservation. Evolutionary conservation is quantified by looking at the sequence conservation and the 3D regional conservation (i.e. structural conservation of patches) in both target (ANCCON and ANC3DCON) and cognates proteins (CON and 3DCON). To calculate ANCCON and CON values, sequence profiles are derived as described(Fernandez-Fuentes et al. 2007). Next, ANCCON corresponds to conservation scores as calculated by al2co(Pei and Grishin 2001) and the CON variable is the ratio between residues with and al2co scores above 1.0 an the number of cognate residues in the interface. Likewise, the ANC3DCON and 3DCON values are calculate but instead of using al2co scores, the normalize regional conservation scores as defined by Landgrad et al(Landgraf et al. 2001) are used. The last input used by PCRPi is the BE index, which represents the predicted binding energy change upon mutation, i.e. *in silico* Alanine scanning, as calculated using FoldX(Guerois et al. 2002).

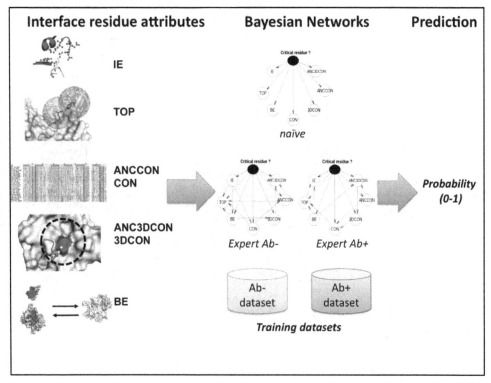

Fig. 3. Overview of the prediction process. PCRPi integrates seven features characterizing interface residues that are used as input variables to three different Bayesian networks, two experts and one naïve, that can be trained with protein complexes including (Ab+) or excluding (Ab-) Antigen-Antibodies complexes. PCRPi outputs a probability where the higher the probability the more likely the residues to be critical, i.e. hot spot residues, for the interaction.

The final part of the prediction is the integration of the data, i.e. IE, TOP, ANCCON, CON, ANC3DCON, 3DCON and BE, into a common probabilistic framework by using BN. PCPRi features three different types BN, two experts and one naïve (Fig. 3). The difference between them is the relationship of dependence between input variables; while naïve BN assumes independence, an expert BN allows conditional dependence between variables (Fig. 3). Both expert and naïve BNs are trained using two specific sets of protein complexes: Ab+ and Ab- (Fig. 3). The Ab+ set corresponds to protein complexes that can include non-evolutionary related complexes such as Antigen-Antibodies complexes while Ab- does not include the latter. The reason being is the lack of sequence conservation in the complementary determining regions of Antibodies, i.e. regions that mediate interaction, which renders evolutionary information meaningless for prediction purposes and thus special BNs were devised to cope with this problem. In terms of performance, PCRPi delivers highly consistent and competitive predictions as shown in the study of the protein complex formed by RAS and VH-HRAS antibody(Tanaka et al. 2007) and a comprehensive comparative study(Assi et al. 2010). Moreover, PCPRi is a central part of a database that compiles and annotates hot spot in protein interfaces: PCRPI-DB (Segura and Fernandez-Fuentes 2011).

Name	Input	Method	URL	Reference
PCRPi	Structure	Machine learning	http://www.bioinsilico.org/PCRPi	(Assi et al. 2010)
Robetta	Structure	Energy-based	http://robetta.bakerlab.org	(Kortemme and Baker 2002)
FoldX	Structure	Energy-based	http://foldx.crg.es	(Guerois et al. 2002)
DrugScorePPi	Structure	Energy-based	http://cpclab.uni-duesseldorf.de/dsppi	(Kruger and Gohlke 2010)
CC/PBSA server	Structure	Energy-based	http://ccpbsa.biologie.uni-erlangen.de/ccpbsa	(Benedix et al. 2009)
KFC	Structure	Machine learning	http://kfc.mitchell-lab.org	(Darnell et al. 2008)
HotPoint	Structure	Scoring function	http://prism.ccbb.ku.edu.tr/hotpoint	(Tuncbag et al. 2009)
ISIS	Sequence	Neural Network	http://rostlab.org/cms/resources/web-services/	(Ofran and Rost 2007)

Table 4. List of online resources for prediction of hot spots.

5. Conclusions and outlook

During the last years, scientists aiming at understanding living organisms at a molecular level have seen their benches become swapped with the sheer amount of information and this burst of data being mirrored by the development of a wide and miscellaneous set of computational tools designed to unveil biologically relevant information from the noisy background. PPIs are among the most crucial events that define the behaviour of a living system and that explains the rise of research efforts and strategies to describe the nature of PPIs. This chapter presents a summary and extensive view on computational methods devoted to predict which proteins participate in PPIs (section 2), which are the regions involved in the interaction (section 3) and which are the most important regions or residues in the interaction (section 4).

In general the prediction tools achieve a high rate of prediction success and are important tools for scientists. However, there are still a number of unmet needs and challenges to be solved. In the case of prediction of PPIs, genome context approaches would benefit from improved definitions of phylogenetic profiles and the masking effect of gene fusion events. Text-mining approaches require further development to reduce false positive rates and increase efficiency. A deeper understanding of the complex interlink between (bio)chemistry, structure and genetics that governs the evolution of protein interfaces would certainly benefit co-evolution-based methods. The correct detection of remote homology between interologs is a major challenge as is the lack of correlation between predicted and observed binding affinities in structure-based methods.

Protein binding site prediction methods have also their own limitations and challenges. The physical forces and chemical properties that drive the interaction between proteins are not fully understood and thus current models do not reflect the binding process accurately. However, the increasing amount of experimental data that is being generated in an important factor that plays in favour of developing novel and more accurate computational tools. Some specific challenges in the field are the prediction of binding sites in proteins that recognize multiple partners (hub proteins) and the distinction between each of the interfaces that are relevant to each of the interacting partners. Current methods cannot properly handle binding events that involve conformational changes in any of the intervening components, including those mediated by intrinsic disordered regions, and thus future efforts need to be directed to tackle this very important question. Finally, the main challenge in the prediction of hot spots is the development of new approaches to bridge the gap between highly computationally expensive methods and those based on simplified models by finding the right balance between the accuracy of the former and the speed of the latter.

6. Acknowledgments

NFF acknowledges support from the Research Councils United Kingdom (RCUK) under the Academic Fellowship scheme. JS acknowledges support from the Leeds Institute of Molecular Medicine (PhD scholarship). JR is supported by a postdoctoral grant awarded by the Consejeria de Educacion y Cultura of the Junta de Comunidades de Castilla La Mancha anb by the European Social Fund. NFF also thanks Dr Gendra for critical reading and insightful comments to the manuscript, and Ms Martina and Ms Daniela G Fernandez for continuing inspiration and motivation. Publication costs were funded by The Biomedical and Health Research Centre (BHRC).

7. References

Airola, A., S. Pyysalo, J. Bjorne, T. Pahikkala, F. Ginter &T. Salakoski (2008). "All-paths graph kernel for protein-protein interaction extraction with evaluation of cross-corpus learning." *BMC Bioinformatics* 9 Suppl 11: S2.

Almlof, M., J. Aqvist, A. O. Smalas &B. O. Brandsdal (2006). "Probing the effect of point mutations at protein-protein interfaces with free energy calculations." *Biophys J* 90(2): 433-42.

Aloy, P., B. Bottcher, H. Ceulemans, C. Leutwein, C. Mellwig, S. Fischer, A. C. Gavin, P. Bork, G. Superti-Furga, L. Serrano &R. B. Russell (2004). "Structure-based assembly of protein complexes in yeast." *Science* 303(5666): 2026-9.

Aloy, P. &R. B. Russell (2003). "InterPreTS: protein Interaction Prediction through Tertiary Structure." *Bioinformatics*. 19(1): 161.

Altschuh, D., A. M. Lesk, A. C. Bloomer &A. Klug (1987). "Correlation of co-ordinated amino acid substitutions with function in viruses related to tobacco mosaic virus." *Journal of Molecular Biology* 193(4): 693-707.

Aranda, B., P. Achuthan, Y. Alam-Faruque, I. Armean, A. Bridge, C. Derow, M. Feuermann, A. T. Ghanbarian, S. Kerrien, J. Khadake, J. Kerssemakers, C. Leroy, M. Menden, M. Michaut, L. Montecchi-Palazzi, S. N. Neuhauser, S. Orchard, V. Perreau, B. Roechert, K. van Eijk &H. Hermjakob (2010). "The IntAct molecular interaction database in 2010." *Nucleic Acids Res* 38(Database issue): D525-31.

Ashkenazy, H., E. Erez, E. Martz, T. Pupko &N. Ben-Tal (2010). "ConSurf 2010: calculating evolutionary conservation in sequence and structure of proteins and nucleic acids." *Nucleic Acids Res* 38(Web Server issue): W529-33.

Assi, S. A., T. Tanaka, T. H. Rabbitts &N. Fernandez-Fuentes (2010). "PCRPi: Presaging Critical Residues in Protein interfaces, a new computational tool to chart hot spots in protein interfaces." *Nucleic Acids Res* 38(6): e86.

Barber, C. B., D. P. Dobkin &H. Huhdanpaa (1996). "The Quickhull algorithm for convex hulls." *ACM TRANSACTIONS ON MATHEMATICAL SOFTWARE* 22(4): 469-483.

Barker, D. &M. Pagel (2005). "Predicting functional gene links from phylogenetic-statistical analyses of whole genomes." *PLoS Comput Biol* 1(1): e3.

Benedix, A., C. M. Becker, B. L. de Groot, A. Caflisch &R. A. Bockmann (2009). "Predicting free energy changes using structural ensembles." *Nat Methods* 6(1): 3-4.

Blaschke, C., R. Hoffmann, J. C. Oliveros &A. Valencia (2001). "Extracting information automatically from biological literature." *Comp Funct Genomics* 2(5): 310-3.

Bogan, A. A. &K. S. Thorn (1998). "Anatomy of hot spots in protein interfaces." *J Mol Biol* 280(1): 1-9.

Bork, P., L. J. Jensen, C. von Mering, A. K. Ramani, I. Lee &E. M. Marcotte (2004). "Protein interaction networks from yeast to human." *Curr Opin Struct Biol* 14(3): 292-9.

Bradford, J. R., C. J. Needham, A. J. Bulpitt &D. R. Westhead (2006). "Insights into protein-protein interfaces using a Bayesian network prediction method." *J Mol Biol* 362(2): 365-86.

Bradford, J. R. &D. R. Westhead (2005). "Improved prediction of protein-protein binding sites using a support vector machines approach." *Bioinformatics* 21(8): 1487-94.

Brown, K. R. &I. Jurisica (2007). "Unequal evolutionary conservation of human protein interactions in interologous networks." *Genome Biol* 8(5): R95.

Burger, L. &E. van Nimwegen (2010). "Disentangling direct from indirect co-evolution of residues in protein alignments." *PLoS computational biology* 6(1): e1000633.

Caffrey, D. R., S. Somaroo, J. D. Hughes, J. Mintseris &E. S. Huang (2004). "Are protein-protein interfaces more conserved in sequence than the rest of the protein surface?" *Protein Sci* 13(1): 190-202.

Chang, D. T., Y. Z. Weng, J. H. Lin, M. J. Hwang &Y. J. Oyang (2006). "Protemot: prediction of protein binding sites with automatically extracted geometrical templates." *Nucleic Acids Res* 34(Web Server issue): W303-9.

Chao, J. A., Y. Patskovsky, S. C. Almo &R. H. Singer (2008). "Structural basis for the coevolution of a viral RNA-protein complex." *Nature Structural and Molecular Biology* 15(1): 103-105.

Chatr-aryamontri, A., A. Ceol, L. M. Palazzi, G. Nardelli, M. V. Schneider, L. Castagnoli &G. Cesareni (2007). "MINT: the Molecular INTeraction database." *Nucleic Acids Res* 35(Database issue): D572-4.

Chen, H. &H.-X. Zhou (2005). "Prediction of interface residues in protein-protein complexes by a consensus neural network method: test against NMR data." *Proteins* 61(1): 21-35.

Chen, P. &J. Li (2010). "Sequence-based identification of interface residues by an integrative profile combining hydrophobic and evolutionary information." *BMC Bioinformatics* 11: 402.

Chen, X.-w. &J. C. Jeong (2009). "Sequence-based prediction of protein interaction sites with an integrative method." *Bioinformatics* 25(5): 585-591.

Chen, X. W. &M. Liu (2005). "Prediction of protein-protein interactions using random decision forest framework." *Bioinformatics* 21(24): 4394-400.

Cho, K. I., D. Kim &D. Lee (2009). "A feature-based approach to modeling protein-protein interaction hot spots." *Nucleic Acids Res* 37(8): 2672-87.

Clackson, T. &J. A. Wells (1995). "A hot spot of binding energy in a hormone-receptor interface." *Science* 267(5196): 383-386.

Clarke, N. D. (1995). "Covariation of residues in the homeodomain sequence family." *Protein Science* 4(11): 2269-2278.

Cole, C. &J. Warwicker (2002). "Side-chain conformational entropy at protein-protein interfaces." *Protein Sci* 11(12): 2860-70.

Darnell, S. J., L. LeGault &J. C. Mitchell (2008). "KFC Server: interactive forecasting of protein interaction hot spots." *Nucleic Acids Res* 36(Web Server issue): W265-9.

Darnell, S. J., D. Page &J. C. Mitchell (2007). "An automated decision-tree approach to predicting protein interaction hot spots." *Proteins* 68(4): 813-23.

de Vries, S. J., A. D. J. van Dijk &A. M. J. J. Bonvin (2006). "WHISCY: what information does surface conservation yield? Application to data-driven docking." *Proteins* 63(3): 479-89.

DeLano, W. L. (2002). "Unraveling hot spots in binding interfaces: progress and challenges." *Curr Opin Struct Biol* 12(1): 14-20.

Diller, D. J., C. Humblet, X. Zhang &L. M. Westerhoff (2010). "Computational alanine scanning with linear scaling semiempirical quantum mechanical methods." *Proteins* 78(10): 2329-37.

Dou, T., C. Ji, S. Gu, J. Xu, K. Ying, Y. Xie &Y. Mao (2006). "Co-evolutionary analysis of insulin/insulin like growth factor 1 signal pathway in vertebrate species." *Frontiers in bioscience : a journal and virtual library* 11: 380-8.

Engelen, S., L. A. Trojan, S. Sacquin-Mora, R. Lavery &A. Carbone (2009). "Joint Evolutionary Trees: A Large-Scale Method To Predict Protein Interfaces Based on Sequence Sampling." *PLoS Comput Biol* 5(1): e1000267.

Enright, A. J., I. Iliopoulos, N. C. Kyrpides &C. A. Ouzounis (1999). "Protein interaction maps for complete genomes based on gene fusion events." *Nature* 402(6757): 86-90.

Evguenieva-Hackenberg, E., P. Walter, E. Hochleitner, F. Lottspeich &G. Klug (2003). "An exosome-like complex in Sulfolobus solfataricus." *EMBO Rep* 4(9): 889-93.

Ewing, R. M., P. Chu, F. Elisma, H. Li, P. Taylor, S. Climie, L. McBroom-Cerajewski, M. D. Robinson, L. O'Connor, M. Li, R. Taylor, M. Dharsee, Y. Ho, A. Heilbut, L. Moore,

S. Zhang, O. Ornatsky, Y. V. Bukhman, M. Ethier, Y. Sheng, J. Vasilescu, M. Abu-Farha, J. P. Lambert, H. S. Duewel, Stewart, II, B. Kuehl, K. Hogue, K. Colwill, K. Gladwish, B. Muskat, R. Kinach, S. L. Adams, M. F. Moran, G. B. Morin, T. Topaloglou &D. Figeys (2007). "Large-scale mapping of human protein-protein interactions by mass spectrometry." *Mol Syst Biol* 3: 89.

Fares, M. A. &D. McNally (2006). "CAPS: Coevolution analysis using protein sequences." *Bioinformatics* 22(22): 2821-2822.

Fariselli, P., F. Pazos, A. Valencia &R. Casadio (2002). "Prediction of protein--protein interaction sites in heterocomplexes with neural networks." *Eur J Biochem* 269(5): 1356-61.

Fernandez-Fuentes, N., B. K. Rai, C. J. Madrid-Aliste, J. E. Fajardo &A. Fiser (2007). "Comparative protein structure modeling by combining multiple templates and optimizing sequence-to-structure alignments." *Bioinformatics* 23(19): 2558-65.

Finn, R. D., M. Marshall &A. Bateman (2005). "iPfam: visualization of protein-protein interactions in PDB at domain and amino acid resolutions." *Bioinformatics* 21(3): 410-2.

Finn, R. D., J. Mistry, B. Schuster-B√∂ckler, S. Griffiths-Jones, V. Hollich, T. Lassmann, S. Moxon, M. Marshall, A. Khanna, R. Durbin, S. R. Eddy, E. L. L. Sonnhammer &A. Bateman (2006). "Pfam: clans, web tools and services." *Nucleic Acids Res* 34(Database issue): D247-51.

Fleishman, S. J., S. D. Khare, N. Koga &D. Baker (2011). "Restricted sidechain plasticity in the structures of native proteins and complexes." *Protein Sci* 20(4): 753-7.

Fu, W., B. E. Sanders-Beer, K. S. Katz, D. R. Maglott, K. D. Pruitt &R. G. Ptak (2009). "Human immunodeficiency virus type 1, human protein interaction database at NCBI." *Nucleic Acids Res* 37(Database issue): D417-22.

Fundel, K., R. Kuffner &R. Zimmer (2007). "RelEx--relation extraction using dependency parse trees." *Bioinformatics* 23(3): 365-71.

Gao, H., Y. Dou, J. Yang &J. Wang (2011). "New methods to measure residues coevolution in proteins." *BMC Bioinformatics* 12: 206.

Glaser, F., D. M. Steinberg, I. A. Vakser &N. Ben Tal (2001). "Residue frequencies and pairing preferences at protein-protein interfaces." *Proteins* 43(2): 89.

Gloor, G. B., L. C. Martin, L. M. Wahl &S. D. Dunn (2005). "Mutual information in protein multiple sequence alignments reveals two classes of coevolving positions." *Biochemistry* 44(19): 7156-7165.

Gobel, U., C. Sander, R. Schneider &A. Valencia (1994). "Correlated mutations and residue contacts in proteins." *Proteins: Structure, Function and Genetics* 18(4): 309-317.

Goh, C.-S. &F. E. Cohen (2002). "Co-evolutionary analysis reveals insights into protein-protein interactions." *Journal of Molecular Biology* 324(1): 177-192.

Goll, J., S. V. Rajagopala, S. C. Shiau, H. Wu, B. T. Lamb &P. Uetz (2008). "MPIDB: the microbial protein interaction database." *Bioinformatics* 24(15): 1743-4.

Grishin, N. V. &M. A. Phillips (1994). "The subunit interfaces of oligomeric enzymes are conserved to a similar extent to the overall protein sequences." *Protein Sci* 3(12): 2455-8.

Grosdidier, S. &J. Fernandez-Recio (2008). "Identification of hot-spot residues in protein-protein interactions by computational docking." *BMC Bioinformatics* 9: 447.

Guerois, R., J. E. Nielsen &L. Serrano (2002). "Predicting changes in the stability of proteins and protein complexes: a study of more than 1000 mutations." *J Mol Biol* 320(2): 369-87.

Guharoy, M. &P. Chakrabarti (2005). "Conservation and relative importance of residues across protein-protein interfaces." *Proc Natl Acad Sci U S A* 102(43): 15447-52.

Hakes, L., S. C. Lovell, S. G. Oliver &D. L. Robertson (2007). "Specificity in protein interactions and its relationship with sequence diversity and coevolution." *Proceedings of the National Academy of Sciences of the United States of America* 104(19): 7999-8004.

Hoffmann, R. &A. Valencia (2004). "A gene network for navigating the literature." *Nat Genet* 36(7): 664.

Hsu, C. M., C. Y. Chen, B. J. Liu, C. C. Huang, M. H. Laio, C. C. Lin &T. L. Wu (2007). "Identification of hot regions in protein-protein interactions by sequential pattern mining." *BMC Bioinformatics* 8 Suppl 5: S8.

Hu, Z., B. Ma, H. Wolfson &R. Nussinov (2000). "Conservation of polar residues as hot spots at protein interfaces." *Proteins* 39(4): 331-42.

Huang, T. W., C. Y. Lin &C. Y. Kao (2007). "Reconstruction of human protein interolog network using evolutionary conserved network." *BMC Bioinformatics* 8: 152.

Huang, Y. J., D. Hang, L. J. Lu, L. Tong, M. B. Gerstein &G. T. Montelione (2008). "Targeting the human cancer pathway protein interaction network by structural genomics." *Mol Cell Proteomics* 7(10): 2048-60.

Hubner, N. C., A. W. Bird, J. Cox, B. Splettstoesser, P. Bandilla, I. Poser, A. Hyman &M. Mann (2010). "Quantitative proteomics combined with BAC TransgeneOmics reveals in vivo protein interactions." *J Cell Biol* 189(4): 739-54.

Hue, M., M. Riffle, J. P. Vert &W. S. Noble (2010). "Large-scale prediction of protein-protein interactions from structures." *BMC Bioinformatics* 11: 144.

Itzhaki, Z., E. Akiva, Y. Altuvia &H. Margalit (2006). "Evolutionary conservation of domain-domain interactions." *Genome Biol* 7(12): R125.

Jones, S. &J. M. Thornton (1997). "Prediction of protein-protein interaction sites using patch analysis." *J.Mol.Biol.* 272(1): 133.

Jonsson, P. F. &P. A. Bates (2006). "Global topological features of cancer proteins in the human interactome." *Bioinformatics* 22(18): 2291-7.

Jothi, R., P. F. Cherukuri, A. Tasneem &T. M. Przytycka (2006). "Co-evolutionary analysis of domains in interacting proteins reveals insights into domain-domain interactions mediating protein-protein interactions." *J Mol Biol* 362(4): 861-75.

Juan, D., F. Pazos &A. Valencia (2008). "High-confidence prediction of global interactomes based on genome-wide coevolutionary networks." *Proc Natl Acad Sci U S A* 105(3): 934-9.

Kastritis, P. L. &A. M. Bonvin (2010). "Are scoring functions in protein-protein docking ready to predict interactomes? Clues from a novel binding affinity benchmark." *J Proteome Res* 9(5): 2216-25.

Kawashima, S., P. Pokarowski, M. Pokarowska, A. Kolinski, T. Katayama &M. Kanehisa (2008). "AAindex: amino acid index database, progress report 2008." *Nucleic Acids Res* 36(Database issue): D202-5.

Kelley, R. &T. Ideker (2005). "Systematic interpretation of genetic interactions using protein networks." *Nat Biotechnol* 23(5): 561-6.

Keskin, O., B. Ma &R. Nussinov (2005). "Hot regions in protein--protein interactions: the organization and contribution of structurally conserved hot spot residues." *J Mol Biol* 345(5): 1281-94.

Kiemer, L. &G. Cesareni (2007). "Comparative interactomics: comparing apples and pears?" *Trends Biotechnol* 25(10): 448-54.

Kim, S., J. Yoon &J. Yang (2008). "Kernel approaches for genic interaction extraction." *Bioinformatics* 24(1): 118-26.

Kim, W. K., J. Park &J. K. Suh (2002). "Large scale statistical prediction of protein-protein interaction by potentially interacting domain (PID) pair." *Genome Inform* 13: 42-50.

Koike, A. &T. Takagi (2004). "Prediction of protein-protein interaction sites using support vector machines." *Protein Engineering Design and Selection* 17(2): 165-173.

Koonin, E. V., Y. I. Wolf &L. Aravind (2001). "Prediction of the archaeal exosome and its connections with the proteasome and the translation and transcription machineries by a comparative-genomic approach." *Genome Res* 11(2): 240-52.

Kortemme, T. &D. Baker (2002). "A simple physical model for binding energy hot spots in protein-protein complexes." *Proc Natl Acad Sci U S A* 99(22): 14116-21.

Kruger, D. M. &H. Gohlke (2010). "DrugScorePPI webserver: fast and accurate in silico alanine scanning for scoring protein-protein interactions." *Nucleic Acids Res* 38(Web Server issue): W480-6.

Kundrotas, P. J., Z. Zhu &I. A. Vakser (2010). "GWIDD: Genome-wide protein docking database." *Nucleic Acids Res* 38(Database issue): D513-7.

Labedan, B., Y. Xu, D. G. Naumoff &N. Glansdorff (2004). "Using quaternary structures to assess the evolutionary history of proteins: the case of the aspartate carbamoyltransferase." *Molecular biology and evolution* 21(2): 364-73.

Lafont, V., M. Schaefer, R. H. Stote, D. Altschuh &A. Dejaegere (2007). "Protein-protein recognition and interaction hot spots in an antigen-antibody complex: free energy decomposition identifies "efficient amino acids"." *Proteins* 67(2): 418-34.

Landgraf, R., I. Xenarios &D. Eisenberg (2001). "Three-dimensional cluster analysis identifies interfaces and functional residue clusters in proteins." *J.Mol.Biol.* 307(5): 1487.

Landon, M. R., D. R. Lancia, Jr., J. Yu, S. C. Thiel &S. Vajda (2007). "Identification of hot spots within druggable binding regions by computational solvent mapping of proteins." *J Med Chem* 50(6): 1231-40.

Lee, H., M. Deng, F. Sun &T. Chen (2006). "An integrated approach to the prediction of domain-domain interactions." *BMC Bioinformatics* 7: 269.

Lewis, A. C. F., R. Saeed &C. M. Deane (2010). "Predicting protein-protein interactions in the context of protein evolution." *Molecular BioSystems* 6: 55-64.

Li, L., B. Zhao, Z. Cui, J. Gan, M. K. Sakharkar &P. Kangueane (2006). "Identification of hot spot residues at protein-protein interface." *Bioinformation* 1(4): 121-6.

Li, X., O. Keskin, B. Ma, R. Nussinov &J. Liang (2004). "Protein-protein interactions: hot spots and structurally conserved residues often locate in complemented pockets that pre-organized in the unbound states: implications for docking." *J Mol Biol* 344(3): 781-95.

Li, Z., L. Wong &J. Li (2011). "DBAC: a simple prediction method for protein binding hot spots based on burial levels and deeply buried atomic contacts." *BMC Syst Biol* 5 Suppl 1: S5.

Liang, S., C. Zhang, S. Liu &Y. Zhou (2006). "Protein binding site prediction using an empirical scoring function." *Nucleic Acids Res* 34(13): 3698-707.

Lichtarge, O., H. R. Bourne &F. E. Cohen (1996). "An evolutionary trace method defines binding surfaces common to protein families." *J.Mol.Biol.* 257(2): 342.

Lise, S., C. Archambeau, M. Pontil &D. T. Jones (2009). "Prediction of hot spot residues at protein-protein interfaces by combining machine learning and energy-based methods." *BMC Bioinformatics* 10: 365.

Lo Conte, L., C. Chothia &J. Janin (1999). "The atomic structure of protein-protein recognition sites." *J.Mol.Biol.* 285(5): 2177.

Lovell, S. C. &D. L. Robertson (2010). "An integrated view of molecular coevolution in protein-protein interactions." *Molecular Biology and Evolution* 27(11): 2567-2575.

Lu, L., H. Lu &J. Skolnick (2002). "MULTIPROSPECTOR: an algorithm for the prediction of protein-protein interactions by multimeric threading." *Proteins* 49(3): 350-64.

Luo, Q., P. Pagel, B. Vilne &D. Frishman (2011). "DIMA 3.0: Domain Interaction Map." *Nucleic Acids Res* 39(Database issue): D724-9.

Ma, B. &R. Nussinov (2007). "Trp/Met/Phe hot spots in protein-protein interactions: potential targets in drug design." *Curr Top Med Chem* 7(10): 999-1005.

Madaoui, H. &R. Guerois (2008). "Coevolution at protein complex interfaces can be detected by the complementarity trace with important impact for predictive docking." *Proceedings of the National Academy of Sciences of the United States of America* 105(22): 7708-7713.

Marcotte, E. M., M. Pellegrini, H. L. Ng, D. W. Rice, T. O. Yeates &D. Eisenberg (1999). "Detecting protein function and protein-protein interactions from genome sequences." *Science* 285(5428): 751-3.

Matthews, L. R., P. Vaglio, J. Reboul, H. Ge, B. P. Davis, J. Garrels, S. Vincent &M. Vidal (2001). "Identification of potential interaction networks using sequence-based searches for conserved protein-protein interactions or "interologs"." *Genome Res* 11(12): 2120-6.

McPartland, J. M., R. W. Norris &C. W. Kilpatrick (2007). "Coevolution between cannabinoid receptors and endocannabinoid ligands." *Gene* 397(1-2): 126-35.

Mika, S. &B. Rost (2006). "Protein-protein interactions more conserved within species than across species." *PLoS Comput Biol* 2(7): e79.

Mintseris, J. &Z. Weng (2005). "Structure, function, and evolution of transient and obligate protein-protein interactions." *Proceedings of the National Academy of Sciences of the United States of America* 102(31): 10930-10935.

Miwa, M., R. Saetre, Y. Miyao &J. Tsujii (2009). "Protein-protein interaction extraction by leveraging multiple kernels and parsers." *Int J Med Inform* 78(12): e39-46.

Moreira, I. S., P. A. Fernandes &M. J. Ramos (2007). "Computational alanine scanning mutagenesis--an improved methodological approach." *J Comput Chem* 28(3): 644-54.

Murakami, Y. &S. Jones (2006). "SHARP2: protein-protein interaction predictions using patch analysis." *Bioinformatics* 22(14): 1794-5.

Murakami, Y. &K. Mizuguchi (2010). "Applying the Naive Bayes classifier with kernel density estimation to the prediction of protein-protein interaction sites." *Bioinformatics* 26(15): 1841-8.

Murali, T., S. Pacifico, J. Yu, S. Guest, G. G. Roberts, 3rd &R. L. Finley, Jr. (2011). "DroID 2011: a comprehensive, integrated resource for protein, transcription factor, RNA and gene interactions for Drosophila." *Nucleic Acids Res* 39(Database issue): D736-43.

Negi, S. S., C. H. Schein, N. Oezguen, T. D. Power &W. Braun (2007). "InterProSurf: a web server for predicting interacting sites on protein surfaces." *Bioinformatics* 23(24): 3397-9.

Neuvirth, H., R. Raz &G. Schreiber (2004). "ProMate: a structure based prediction program to identify the location of protein-protein binding sites." *J Mol Biol* 338(1): 181-99.

Ochoa, D. &F. Pazos (2010). "Studying the co-evolution of protein families with the Mirrortree web server." *Bioinformatics* 26(10): 1370-1.

Ofran, Y. &B. Rost (2003). "Predicted protein-protein interaction sites from local sequence information." *FEBS Lett* 544(1-3): 236-9.

Ofran, Y. &B. Rost (2007). "ISIS: interaction sites identified from sequence." *Bioinformatics* 23(2): e13-6.

Ofran, Y. &B. Rost (2007). "Protein-protein interaction hotspots carved into sequences." *PLoS Comput Biol* 3(7): e119.

Pagel, P., P. Wong &D. Frishman (2004). "A domain interaction map based on phylogenetic profiling." *J Mol Biol* 344(5): 1331-46.

Panni, S., L. Dente &G. Cesareni (2002). "In vitro evolution of recognition specificity mediated by SH3 domains reveals target recognition rules." *J Biol Chem* 277(24): 21666-74.

Park, D., R. Singh, M. Baym, C. S. Liao &B. Berger (2011). "IsoBase: a database of functionally related proteins across PPI networks." *Nucleic Acids Res* 39(Database issue): D295-300.

Patil, A., K. Nakai &H. Nakamura (2011). "HitPredict: a database of quality assessed protein-protein interactions in nine species." *Nucleic Acids Res* 39(Database issue): D744-9.

Pazos, F. &A. Valencia (2008). "Protein co-evolution, co-adaptation and interactions." *Embo J* 27(20): 2648-55.

Pei, J. &N. V. Grishin (2001). "AL2CO: calculation of positional conservation in a protein sequence alignment." *Bioinformatics* 17(8): 700-12.

Pellegrini, M., E. M. Marcotte, M. J. Thompson, D. Eisenberg &T. O. Yeates (1999). "Assigning protein functions by comparative genome analysis: protein phylogenetic profiles." *Proc.Natl.Acad.Sci.U.S.A* 96(8): 4285.

Pesquita, C., D. Faria, A. O. Falcao, P. Lord &F. M. Couto (2009). "Semantic similarity in biomedical ontologies." *PLoS Comput Biol* 5(7): e1000443.

Porollo, A. &J. Ç. Meller (2007). "Prediction-based fingerprints of protein-protein interactions." *Proteins* 66(3): 630-45.

Prasad, T. S., K. Kandasamy &A. Pandey (2009). "Human Protein Reference Database and Human Proteinpedia as discovery tools for systems biology." *Methods Mol Biol* 577: 67-79.

Presser, A., M. B. Elowitz, M. Kellis &R. Kishony (2008). "The evolutionary dynamics of the Saccharomyces cerevisiae protein interaction network after duplication." *Proceedings of the National Academy of Sciences of the United States of America* 105(3): 950-954.

Qin, S. &H. X. Zhou (2007). "meta-PPISP: a meta web server for protein-protein interaction site prediction." *Bioinformatics* 23(24): 3386-7.

Res, I., I. Mihalek &O. Lichtarge (2005). "An evolution based classifier for prediction of protein interfaces without using protein structures." *Bioinformatics* 21(10): 2496-2501.

Riley, R., C. Lee, C. Sabatti &D. Eisenberg (2005). "Inferring protein domain interactions from databases of interacting proteins." *Genome Biol* 6(10): R89.

Rodionov, A., A. Bezginov, J. Rose &E. R. Tillier (2011). "A new, fast algorithm for detecting protein coevolution using maximum compatible cliques." *Algorithms for molecular biology : AMB* 6: 17.

Rual, J. F., K. Venkatesan, T. Hao, T. Hirozane-Kishikawa, A. Dricot, N. Li, G. F. Berriz, F. D. Gibbons, M. Dreze, N. Ayivi-Guedehoussou, N. Klitgord, C. Simon, M. Boxem, S. Milstein, J. Rosenberg, D. S. Goldberg, L. V. Zhang, S. L. Wong, G. Franklin, S. Li, J. S. Albala, J. Lim, C. Fraughton, E. Llamosas, S. Cevik, C. Bex, P. Lamesch, R. S. Sikorski, J. Vandenhaute, H. Y. Zoghbi, A. Smolyar, S. Bosak, R. Sequerra, L. Doucette-Stamm, M. E. Cusick, D. E. Hill, F. P. Roth &M. Vidal (2005). "Towards a proteome-scale map of the human protein-protein interaction network." *Nature* 437(7062): 1173-8.

Salwinski, L., C. Miller, A. Smith, F. Pettit, J. Bowie &D. Eisenberg (2004). "The Database of Interacting Proteins: 2004 update." *Nucleic Acids Res* 32: D449 - D451.

Segura, J. &N. Fernandez-Fuentes (2011). "PCRPi-DB: a database of computationally annotated hot spots in protein interfaces." *Nucleic Acids Res* 39(Database issue): D755-60.

Segura, J., P. F. Jones &N. Fernandez-Fuentes (2011). "Improving the prediction of protein binding sites by combining heterogeneous data and Voronoi diagrams." *BMC Bioinformatics* 12: 352.

Shoemaker, B. &A. Panchenko (2007). "Deciphering protein-protein interactions. Part II. Computational methods to predict protein and domain interaction partners." *PLoS Comput Biol* 3(4): 595 - 601.

Shoemaker, B. A. &A. R. Panchenko (2007). "Deciphering protein-protein interactions. Part II. Computational methods to predict protein and domain interaction partners." *PLoS Comput Biol* 3(4): e43.

Shoemaker, B. A., D. Zhang, R. R. Thangudu, M. Tyagi, J. H. Fong, A. Marchler-Bauer, S. H. Bryant, T. Madej &A. R. Panchenko (2010). "Inferred Biomolecular Interaction Server--a web server to analyze and predict protein interacting partners and binding sites." *Nucleic Acids Res* 38(Database issue): D518-24.

Sikic, M., S. Tomic &K. Vlahovicek (2009). "Prediction of protein-protein interaction sites in sequences and 3D structures by random forests." *PLoS Comput Biol* 5(1): e1000278.

Stark, C., B. J. Breitkreutz, A. Chatr-Aryamontri, L. Boucher, R. Oughtred, M. S. Livstone, J. Nixon, K. Van Auken, X. Wang, X. Shi, T. Reguly, J. M. Rust, A. Winter, K. Dolinski &M. Tyers (2011). "The BioGRID Interaction Database: 2011 update." *Nucleic Acids Res* 39(Database issue): D698-704.

Stein, A., A. Ceol &P. Aloy (2011). "3did: identification and classification of domain-based interactions of known three-dimensional structure." *Nucleic Acids Res* 39(Database issue): D718-23.

Stein, A., R. Mosca &P. Aloy (2011). "Three-dimensional modeling of protein interactions and complexes is going -omics." *Current Opinion in Structural Biology* 21(2): 200-208.

Stelzl, U. &E. E. Wanker (2006). "The value of high quality protein-protein interaction networks for systems biology." *Curr Opin Chem Biol* 10(6): 551-8.

Stelzl, U., U. Worm, M. Lalowski, C. Haenig, F. H. Brembeck, H. Goehler, M. Stroedicke, M. Zenkner, A. Schoenherr, S. Koeppen, J. Timm, S. Mintzlaff, C. Abraham, N. Bock, S. Kietzmann, A. Goedde, E. Toksoz, A. Droege, S. Krobitsch, B. Korn, W. Birchmeier, H. Lehrach &E. E. Wanker (2005). "A human protein-protein interaction network: a resource for annotating the proteome." *Cell* 122(6): 957-68.

Stumpf, M., T. Thorne, E. de Silva, R. Stewart, H. An, M. Lappe &C. Wiuf (2008). "Estimating the size of the human interactome." *Proceedings of the National Academy of Sciences of the United States of America* 105(19): 6959 - 6964.

Swarbreck, D., C. Wilks, P. Lamesch, T. Z. Berardini, M. Garcia-Hernandez, H. Foerster, D. Li, T. Meyer, R. Muller, L. Ploetz, A. Radenbaugh, S. Singh, V. Swing, C. Tissier, P. Zhang &E. Huala (2008). "The Arabidopsis Information Resource (TAIR): gene structure and function annotation." *Nucleic Acids Res* 36(Database issue): D1009-14.

Szklarczyk, D., A. Franceschini, M. Kuhn, M. Simonovic, A. Roth, P. Minguez, T. Doerks, M. Stark, J. Muller, P. Bork, L. J. Jensen &C. von Mering (2011). "The STRING database in 2011: functional interaction networks of proteins, globally integrated and scored." *Nucleic Acids Res* 39(Database issue): D561-8.

Ta, H. X., P. Koskinen &L. Holm (2011). "A novel method for assigning functional linkages to proteins using enhanced phylogenetic trees." *Bioinformatics* 27(5): 700-6.

Tanaka, T., R. L. Williams &T. H. Rabbitts (2007). "Tumour prevention by a single antibody domain targeting the interaction of signal transduction proteins with RAS." *EMBO J* 26(13): 3250-9.

Tikk, D., P. Thomas, P. Palaga, J. Hakenberg &U. Leser (2010). "A comprehensive benchmark of kernel methods to extract protein-protein interactions from literature." *PLoS Comput Biol* 6: e1000837.

Tillier, E. R. M. &R. L. Charlebois (2009). "The human protein coevolution network." *Genome Research* 19(10): 1861 -1871.

Travers, S. A. A. &M. A. Fares (2007). "Functional coevolutionary networks of the Hsp70-Hop-Hsp90 system revealed through computational analyses." *Molecular Biology and Evolution* 24(4): 1032-1044.

Tuncbag, N., A. Gursoy &O. Keskin (2009). "Identification of computational hot spots in protein interfaces: combining solvent accessibility and inter-residue potentials improves the accuracy." *Bioinformatics* 25(12): 1513-20.

Tuncbag, N., A. Gursoy, R. Nussinov &O. Keskin (2011). "Predicting protein-protein interactions on a proteome scale by matching evolutionary and structural similarities at interfaces using PRISM." *Nat. Protocols* 6(9): 1341-1354.

Venkatesan, K., J. F. Rual, A. Vazquez, U. Stelzl, I. Lemmens, T. Hirozane-Kishikawa, T. Hao, M. Zenkner, X. Xin, K. I. Goh, M. A. Yildirim, N. Simonis, K. Heinzmann, F. Gebreab, J. M. Sahalie, S. Cevik, C. Simon, A. S. de Smet, E. Dann, A. Smolyar, A. Vinayagam, H. Yu, D. Szeto, H. Borick, A. Dricot, N. Klitgord, R. R. Murray, C. Lin, M. Lalowski, J. Timm, K. Rau, C. Boone, P. Braun, M. E. Cusick, F. P. Roth, D. E. Hill, J. Tavernier, E. E. Wanker, A. L. Barabasi &M. Vidal (2009). "An empirical framework for binary interactome mapping." *Nat Methods* 6(1): 83-90.

Vlahovicek, K., A. Pintar, L. Parthasarathi, O. Carugo &S. Pongor (2005). "CX, DPX and PRIDE: WWW servers for the analysis and comparison of protein 3D structures." *Nucleic Acids Res* 33(Web Server issue): W252-4.

Walhout, A. J., R. Sordella, X. Lu, J. L. Hartley, G. F. Temple, M. A. Brasch, N. Thierry-Mieg &M. Vidal (2000). "Protein interaction mapping in C. elegans using proteins involved in vulval development." *Science* 287(5450): 116-22.

Wang, B., P. Chen, D.-S. Huang, J.-j. Li, T.-M. Lok &M. R. Lyu (2006). "Predicting protein interaction sites from residue spatial sequence profile and evolution rate." *FEBS Lett* 580(2): 380-4.

Wass, M. N., G. Fuentes, C. Pons, F. Pazos &A. Valencia (2011). "Towards the prediction of protein interaction partners using physical docking." *Mol Syst Biol* 7: 469.

Williams, S. G. &S. C. Lovell (2009). "The effect of sequence evolution on protein structural divergence." *Molecular Biology and Evolution* 26(5): 1055-1065.

Wu, J., S. Kasif &C. DeLisi (2003). "Identification of functional links between genes using phylogenetic profiles." *Bioinformatics* 19(12): 1524-30.

Xia, J. F., X. M. Zhao, J. Song &D. S. Huang (2010). "APIS: accurate prediction of hot spots in protein interfaces by combining protrusion index with solvent accessibility." *BMC Bioinformatics* 11: 174.

Yanai, I., A. Derti &C. DeLisi (2001). "Genes linked by fusion events are generally of the same functional category: a systematic analysis of 30 microbial genomes." *Proc Natl Acad Sci U S A* 98(14): 7940-5.

Yang, Z., Y. Lin, J. Wu, N. Tang, H. Lin &Y. Li (2011). "Ranking support vector machine for multiple kernels output combination in protein-protein interaction extraction from biomedical literature." *Proteomics* 11(19): 3811-7.

Yellaboina, S., A. Tasneem, D. V. Zaykin, B. Raghavachari &R. Jothi (2011). "DOMINE: a comprehensive collection of known and predicted domain-domain interactions." *Nucleic Acids Res* 39(Database issue): D730-5.

Yogurtcu, O. N., S. B. Erdemli, R. Nussinov, M. Turkay &O. Keskin (2008). "Restricted mobility of conserved residues in protein-protein interfaces in molecular simulations." *Biophys J* 94(9): 3475-85.

Yu, C.-Y., L.-C. Chou &D. Chang (2010). "Predicting protein-protein interactions in unbalanced data using the primary structure of proteins." *BMC Bioinformatics* 11(1): 167.

Yuan, Z., J. Zhao &Z.-X. Wang (2003). "Flexibility analysis of enzyme active sites by crystallographic temperature factors." *Protein Eng* 16(2): 109-14.

Zhang, Y., H. Lin, Z. Yang &Y. Li (2011). "Neighborhood hash graph kernel for protein-protein interaction extraction." *J Biomed Inform.*

Zhou, H. X. &Y. Shan (2001). "Prediction of protein interaction sites from sequence profile and residue neighbor list." *Proteins* 44(3): 336-43.

7

Scalable, Integrative Analysis and Visualization of Protein Interactions

David Otasek[1], Chiara Pastrello[2,1] and Igor Jurisica[1,3]

[1]*Ontario Cancer Institute the Campbell Family Institute for Cancer Research,*
University Health Network, Toronto, Ontario,
[2]*CRO Aviano, National Cancer Institute, Aviano,*
[3]*Department of Computer Science and Medical Biophysics,*
University of Toronto, Toronto,
[1,3]*Canada*
[2]*Italy*

1. Introduction

Biology offers a diversity of problems, leading to many computational biology workflows, including tasks where network visualization is helpful to interpret and analyse data. High-throughput screening techniques generate large amounts of data useful for the comprehension of the biological mechanisms underlying different diseases. The need for agile tools to handle such data and analyse it correctly has become continuously more evident.

Individual network visualization systems differ greatly in terms of the features and standards they support, and consequently the analyses they enable. Importantly, users have a broad range of skills and expectations, ranging from biology to computational biology. As a result, network visualization tools must satisfy diverse requirements and thus offer different user interfaces and features. In this role, they are also fundamental in helping scientists in different fields integrate their knowledge and their data in an interdisciplinary approach to research.

The number of '-omics' disciplines that use high-throughput techniques and that can benefit from a network approach are increasing. The diverse data that can be represented as a graph includes physical protein-protein interactions (PPIs), metabolic networks (Swainston et al., 2011), genetic co-expression (Helaers et al., 2011), gene regulatory networks (Longabaugh, 2012), microRNA-target (Shirdel et al., 2011) and drug-target associations (Morrow et al., 2010). In this chapter we focus on physical PPIs.

Proteins are key players in virtually all biological events that take place within and between cells and often accomplish their function as part of large molecular machines, whose action is coordinated through intricate regulatory networks of transient PPIs. The understanding of the interrelationships between molecules is the basis for an understanding of the behaviour of biological systems (Stein et al., 2011).

The analysis of the full proteome is possible with techniques such as mass spectrometry and protein microarray, which can be integrated with targeted approaches such as yeast-2-hybrid screen, immune precipitation and affinity purification. So far, PPI discovery methods are not accurate enough to be used alone, but the combination of different techniques can help to build an accurate interactome map (Remmerie et al. 2011). Still, this kind of analysis can only indicate that two proteins interact but does not reveal the molecular details or the mechanism of binding captured in high resolution three-dimensional (3D) structures, in which individual residue contacts are resolved and the interaction interfaces characterized. Moreover, they do not capture transient interactions and post translational modifications (PTMs) that can be addressed by techniques such as immobilized metal affinity chromatography (IMAC) mass spectrometry for protein phosphorylation analysis.

It becomes evident that the analysis of protein interactions is already a huge field with a plethora of data coming from different sources that can be improved by computational techniques and integrative network visualization and analysis. It is even more interesting to integrate PPI data with protein-target interaction data to have a wider view of the environmental context that influences network operations.

In this context, a pathway-centric analysis can help to elucidate the role and the importance of proteins in the context of the cell environment, specifically when the pathways can be related to the process/disease being studied. However, it is mandatory to be aware of the limits of this analysis, due to the cross-talk among pathways: a singular protein, in fact, can be associated or interact with multiple pathways so none of the pathways can be considered a single actor but rather a piece of a bigger puzzle (Kreeger & Lauffenburger, 2010).

Another intriguing aspect that protein-target interactions can describe is the relationship between protein exogenous molecules like drugs or toxins (Yu, 2011). The analysis of networks generated from drug-target and protein-target interactions can highlight different molecules that can be responsible of the response or resistance to a certain drug as well as alternative drugs that can target disease specific proteins.

2. Network visualization tools

There are dozens of applications available for the visualization of biological networks, each with its own focus, work-flow and tools (Pavlopoulos et al., 2008; Gehlenborg et al., 2010). We will describe some of the most common features and workflows involved in using these applications, with brief discussion of NAViGaTOR (Brown et al., 2009; McGuffin and Jurisica, 2009; Djebbari et al., 2011), Cytoscape (Smoot et al., 2011), VANTED (Björn et al., 2006) and VisANT (Mellor et al., 2004), four popular multi-platform biological network visualization applications.

2.1 Biological networks as annotated graphs

The most basic mathematical structure common to all of these applications is the graph, a collection of objects connected by links, referred to as nodes and edges. These objects are abstractions of real-world biological entities, where nodes could represent proteins, genes, molecules, drugs, etc. and edges could represent physical protein-protein, metabolic, or genetic interactions, microRNA to target associations, correlation, similarity relationships, etc. Edges can be directed or undirected, weighted or not. In a case like gene regulation, Gene A may regulate Gene B, but the relationship may not be symmetric, meaning Gene B does not regulate Gene A. These models of biological networks have differing levels of support across various applications. An application may only support a small subset of node and edge types in order to specialize on one particular model, such as VisANT, which integrates many specialized tools for tasks such as Gene Ontology (GO) annotation, name resolution and online searches. Other applications may be more open-ended to provide support for as many models as possible, such as NAViGaTOR, Cytoscape and VANTED. The advantage of such a model is versatility, but it comes at the cost of having to manually define the nature of each node or edge via annotation.

To populate a graph within an application, the application must support one or more input formats. Often, the most basic level of input is either plain text or spreadsheet files such as Excel XLS format. For more graph-specific data, such as layout, GML can be used. To support more complex and structured biological data, several community standards exist: PSI-MI, BioPAX, and SBML.

Adding new nodes and edges to an existing graph can generally be done manually or by adding additional interactions from a supported database or file format. Some applications may have a workspace that supports concurrent, multiple graphs, which can then be combined or compared in various ways. Cytoscape and NAViGaTOR both support this type of workspace.

Once the graph is loaded within an application, a researcher may wish to add additional annotations, such as gene or protein expression, experimental confidence measures or Gene Ontology (The Gene Ontology Consortium, 2000) to their graph objects. Data from in-house sources must generally conform to the application used; generally, this is in the form of spreadsheets or text data with varying degrees of format flexibility. The researcher can also call upon more specialized data from public databases, such as UniProt, Entrez, KEGG or Genbank, either through the import of files or from direct access to the database through the application or a plug-in.

The amount of biological networks available to the researcher is ever expanding, and the size of the networks involved in many types of analysis is in order of thousands of nodes and edges. For example, the yeast interactome comprises 23,918 interactions according to DIP and 152,877 known and predicted interactions in I2D, the Interologous Interaction Database (http://ophid.utoronto.ca/i2d), an integrated database of PPIs from curated databases, experimental sources and predicted interactions (Niu et al., 2010; Brown and Jurisica, 2007; Brown and Jurisica 2005). While the researcher may only be interested in a small portion of the network in question, the scalability of an individual application and its analysis methods to networks of such size can be a considerable advantage.

2.2 Network visualization

Part of the challenge of visualizing a network is the laying out of the graph in a comprehensible manner. For smaller graphs, manual editing of node positions may be sufficient. With the aforementioned instances of graphs in the order of thousands of interactions, more robust tools are available with which to lay out a graph. Automated graph layout algorithms, such as the force-directed and hierarchical, make the process easier, but often produce messy, uninterpretable graphs. Manual control over the placement of nodes and specialized tools for doing so are often necessary, from simple movement of single nodes to alignments in circles and lines to manipulate groups of nodes.

Algorithms for graph analysis are generally included in each application. Here, the number and type of analyses available are wildly variable. Algorithms can be used to find important graph properties, such as node degree, centrality, shortest paths, cliques and clusters. In addition, diverse biology-specific algorithms exist such as GeneMANIA (Montojo et al., 2010). Some applications may be designed specifically for one type of analysis while others contain a variety of analysis methods and in some cases allow for the addition of third party methods through plug-ins (NAViGaTOR, Cytoscape) or scripting languages (VisANT).

How an application chooses to visualize a graph is also variable. Nodes can be represented as anything from basic geometric shapes with variable size, color and transparency to application specific or user supplied bit-mapped images (Cytoscape, VANTED) or even other data visualizations such as bar charts (VANTED, VisANT). Edges can be straight, curved, displayed with various dot or dash schemes and can have variable widths, colors and transparencies. To make certain attributes readily visible, it is also possible in some instances to map an attribute to a visual property, such as color or size. All four of our example applications have different implementations of such mapping; the utility of a specific implementation is dependent upon the needs and competencies of the individual researcher.

Once the graph satisfies the requirements envisioned by the researcher, its state must be stored or exported. Proprietary formats are generally the norm for most programs, as visualization and data are often application specific and must be stored for later editing. Export formats often take the form of community standards (PSI-MI, BioPAX) and graphical exports. Graphical export is generally the final stage before publication. Usually, this can be done in bitmap (JPEG, TIFF, PNG, etc.) or vector (SVG, PDF, etc) formats, the latter being preferable for publication, as it can be resized and manipulated without loss of quality.

2.3 NAViGaTOR

NAViGaTOR (Network Analysis, Visualization and Graphing Toronto; http://ophid.utoronto.ca/navigator) is a network and graph visualization application with an emphasis on large graphs with integrated data (Brown et al., 2009). Data can be imported using diverse formats, ranging from community standards such as PSI-MI XML (Kerrien et al., 2007), BioPAX (Demir et al., 2010) or GML (Himsolt, 1996), to user-defined text files. Though the application is geared towards protein-protein interactions, the graph implementation within NAViGaTOR is not PPI specific, and can be used to model many types of real world or theoretical objects. Nodes and edges can have data associated with them, from simple numeric or text data to structured XML. Once imported, graphs can be combined from within a multi-graph workspace using combinations of cut, copy and paste operations. Additional data for the annotation of existing graphs can be imported using compatible files or online resources, such as I2D, cPath, or the one of the many online databases implementing the PSICQUIC web service.

Graphs generated by the above methods can quickly increase in size to thousands of nodes and edges. NAViGaTOR was designed with networks of this size in mind. While graphs this size do create a demand for both memory and processing power to render, layout and navigate, the conservation of important paths and data is important to end-user analysis, particularly since most graphs of interest are subsets of a much larger interaction networks. NAViGaTOR approaches the problem of limited computing resources through the combination of a powerful OpenGL rendering engine through JOGL, and a suite of efficient layout, search and analysis tools. The JOGL rendering system gives the application access to the graphic processing power of the OpenGL compliant hardware of most graphics cards, allowing the application to use the CPU for more intensive graph operations.

NAViGaTOR supports several layout algorithms tailored for large graphs, including GRIP (Graph Drawing with Intelligent Placement) and several variants of the force directed algorithm. These algorithms come in both single and multi-threaded modes to take advantage of computers with multi-core CPUs.

When the structure and data contained within a graph are sufficient, the user can then interact with the graph, identifying significant nodes, edges or subsets of the graph using a variety of searches, spreadsheet tables and algorithms. Online or file supported databases can also be used to indicate known pathways and complexes within the data.

Users can highlight interesting structures within a graph with a variety of methods. Nodes and edges can be assigned visual properties to differentiate them from each other. Nodes can be given different colors, sizes, and highlighting styles. Edges can be given different colors, widths and styles and have the option to be rendered as user adjustable curves. Transparency can be used on both nodes and edges to either increase or decrease the visibility of graph objects.

The user can save the file in native NAViGaTOR format, GML, PSI-MI or delimited plain text. In addition, for presentation or publication purposes, the graph can be exported to one of several graphical formats, including JPEG, PNG, TIFF, SVG and PDF.

3. Iterative expansion of a protein interaction network

The increasing amount of data that can be collected from high-throughput analyses is accelerating research in the field of molecular biology; however, data of this type is also challenging due to its size. It can be used either for knowledge-based targeted analyses, meaning to improve the understanding of the role of an important well-known player in a specific field of interest (for example of BRCA1 in breast cancer), or unbiased analyses to understand the processes involved in a specific behaviour without a priori knowledge (for example, which genes/proteins are responsible for the poor survival of patients with pancreatic cancer?)

For our example, we have a list of potential interactors for a hypothetical protein of interest, PRO1, generated by computational PPI prediction. Also at our disposal are two meta analyses efforts, specifying the number of ovarian or prostate cancer related studies found in which the gene and its interactors were significantly deregulated. All other data will be collected from publicly available resources, including a PPI database, and a catalogue of drugs and their gene targets.

For our example, we will start with our experimental data in a tabular format. Data such as this can be obtained from any number of sources, from high-throughput experiments to computational predictions. In our case, we have 21,302 predicted PPIs. Our analysis has produced a confidence metric associated with each interaction, ranging from 0 to 1.0. This confidence metric can be used to reduce the number of interactions we are dealing with to a more manageable size by removing lower confidence interactions. Our cut-off for high confidence will be 0.892, a value determined by cross validation. This leaves us with only 39 interactions, a far more manageable number for the next analysis steps. More complex filtering can be done through a simple spreadsheet application, such as Excel, or with a mathematical application such as R or Matlab.

At this point, we translate this data into a pair-wise table of PPIs, and import this table into NAViGaTOR. While NAViGaTOR supports several formats for loading interactions, we have chosen the tab-delimited format to facilitate easy translation from our original data. Other interaction data sets can be imported using community standard file formats, such as BioPAX, GML, PSI-MI XML and PSI-MITAB. Though these formats are harder to construct, they can contain more structured data, and facilitate easier data interchange among diverse programs and databases.

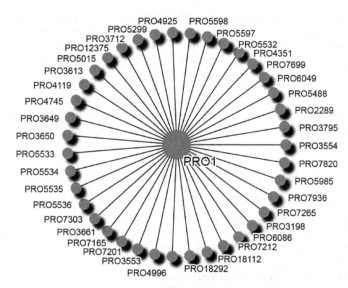

Fig. 1. Example graph containing hypothetical protein PRO1, with interactors loaded from experimental data. Tabular view of the data is available as a supplemental material (http:// http://www.cs.utoronto.ca/~juris/data/intech12/).

Loading our pair-wise data, we get a very basic view (Figure 1). The visualization of this network at this stage is a spoke diagram with PRO1 in the center, and offers little information to the researcher that could not have been seen through a simple spreadsheet. We already have data regarding 39 interactions in the form of the confidence metric imported from our initial study. This can be mapped to one or more visual attributes using NAViGaTORs filter framework. In this case, we can make the highest confidence interactions more visible by applying a filter to map confidence to both edge width and transparency (Figure 2).

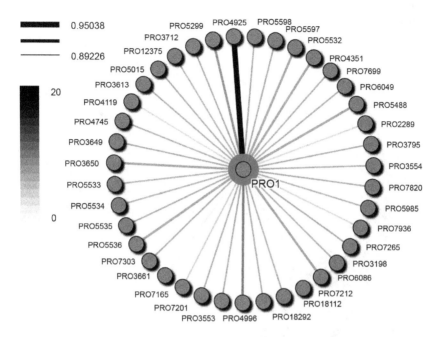

Fig. 2. Example graph with experimental interaction confidence mapped to edge width and transparency.

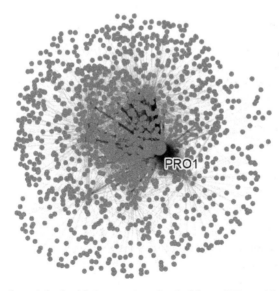

Fig. 3. Example graph enriched with interactions loaded from I2D, and laid out using the GRID algorithm.

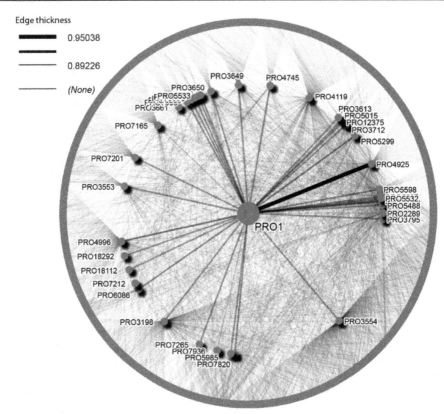

Fig. 4. Example graph laid out hierarchically, with PRO1 in a central position.

This is better, but still not that much more informative. One way of enriching our isolated data is by viewing it in the context of known and predicted interactions. I2D, the Interologous Interaction Database (http://ophid.utoronto.ca/i2d; (Brown et al., 2005, Brown et al., 2007)), will be our source for these interactions. NAViGaTOR offers an I2D plug-in, which enables the researcher to easily add interactions to the existing graph. NAViGaTOR also has the PSICQUIC search plug-in, which supports the searching of databases that implement the PSICQUIC interface (Aranda et al., 2011). To further support the openness and versatility of PPI integration, NAViGaTOR can import additional interactions from the same file formats listed above. If a database does not support any of these formats, finding or building a representation of the database in tab-delimited format may be an option as well. Our interaction search returns 1,367 nodes and 3,192 edges (Figure 3).

At this point, the graph has become more complex, and the force-directed layout is not helpful in interpreting it. Several options exist at this point for manually laying out objects in the graph. The user can select 'fix' nodes within the graph and either move them manually (which would be very labor intensive and inflexible) or lay them out with an array of tools such as linear, circular, arc or radial layout. We will use the radial layout method, starting with PRO1 as our central node and extending to a depth of 2. This gives us a hierarchical arrangement of nodes starting with PRO1 in the centre, with its immediate interactors arranged circularly around it, and their interactors in turn arranged around them (Figure 4).

3.1 Ambiguity of protein names

When combining data from different sources, the users' choice of protein nomenclature becomes extremely important. Although a researcher knows which genes or proteins they are referring to, queries to a database require additional levels of specificity to resolve ambiguities in entity names.

For example, DLC1 has the following SwissProt identifiers: Q96QB1, Q9Y238, P63167, Q7Z5R8, Q45XF9, Q86UC6. However, names in literature could be ambiguous and confusing, potentially resulting in incorrect interpretation and analyses:

- **DLC1** (ARHGAP7) (KIAA1723) (STARD12) [**Rho GTPase-activating protein 7** (Rho-type GTPase-activating protein 7) (Deleted in liver cancer 1 protein) (Dlc-1) (StAR-related lipid transfer protein 12) (START domain-containing protein 12) (StARD12) (HP protein)]
- **DLEC1** (DLC1) [**Deleted in lung and esophageal cancer protein 1** (Deleted in lung cancer protein 1) (DLC-1)]
- **DYNLL1** (DLC1) (DNCL1) (DNCLC1) (HDLC1) [**Dynein light chain 1, cytoplasmic** (Dynein light chain LC8-type 1) (8 kDa dynein light chain) (DLC8) (Protein inhibitor of neuronal nitric oxide synthase) (PIN)]
- **DLC1** [**Deleted in liver cancer 1** variant 2 (Fragment)]
- **DLC1** [DLC1 protein]

Similarly, many papers refer to SHC – but details about which variant and which species are frequently "hidden" in the supplemental information (http://www.cs.utoronto.ca/~juris/data/intech12/). Yet, there are at least four variants in mouse and human. Sometimes, a radical change in nomenclature is required, such as in case of Caspases (Alnemri et al., 1986). Systematic analysis led to redefying various ICE, MACH, MCH genes into Caspase1-10 (Alnemri et al., 1986).

There are many different standards of referring to genes and proteins: UniProt (http://www.uniprot.org) (Jain et al., 2009), Ensembl (http://www.ensembl.org) (Flicek et al., 2012), EBI IPI (http://www.ebi.ac.uk) (Kersey et al., 2004), Gene Cards (http://www.genecards.org) (Safran et al., 2010), NCBI Gene (http://www.ncbi.nlm.nih.gov) (Maglott et al., 2010) are just a few examples of databases that attempt to systematically characterize and describe genes and proteins. Each database has its own focus and strengths, and different interaction or annotation databases may choose any one of these standards to organize their data. In this example, and in many other case uses of NAViGaTOR, the user may have to import data from one or more databases that use different nomenclatures. To facilitate the use of multiple nomenclatures, NAViGaTOR can store multiple IDs per node as a text feature, allowing alternative keys for node identification. When combining data from two or more databases using different formats, the user must translate between these different nomenclatures. This must be done very carefully and methodically, as this additional translation step often effects the data returned. For example, UniProt stores mappings from its own accession IDs to Ensembl

Gene IDs, and Ensembl stores mappings from its own IDs to UniProt. However, respectively, they return 55,639 unique UniProt accession IDs for 20,995 unique Ensembl gene IDs and 21,735 unique Ensembl gene IDs for 63,370 unique UniProt accession IDs. The mapping is clearly different depending on which method is used. There is no definitive mapping available in situations such as these: it is up to the individual researcher to choose and document the translations used to amalgamate their data in a fashion that is replicable. Bearing this in mind during the earlier stages of experiment design will make this process much easier and less prone to confusion or ambiguity.

3.2 Associating data with an existing graph

Though better organized, we still have in excess of 1,000 nodes and 3,000 interactions, and to better identify nodes and edges that represent novel research material, we must associate more data with those objects.

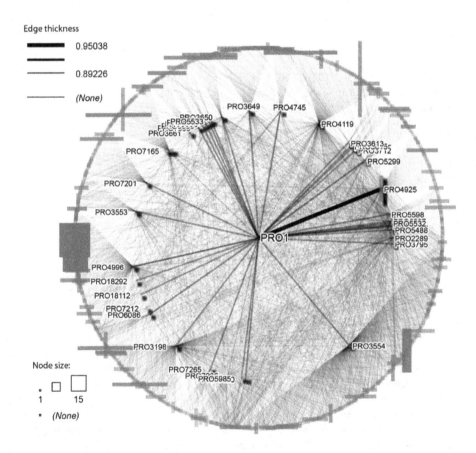

Fig. 5. Example graph with numbers of referencing studies in ovarian and prostate mapped to node width and height.

We can for example integrate PPIs with the gene expression results obtained from our literature studies. Each file contains several values associated with each gene, specifying the number of studies in which the gene was down-regulated, up-regulated and a total representing both (Figure 5). We will also generate a third file representing the total studies in which the gene was found to have been significantly deregulated, which simply sums the totals for the previous two files. Similarly to the opening of the initial experiment, NAViGaTOR requires a unique identifier column to be specified. In this case, because we are only concerned with data to be associated with nodes, the program only requires a single Node ID column. This process is the same for the prostate, ovarian cancer and generated data sets. To visualize this data, we will add another filter, this time mapping the total number of significantly deregulated studies in ovarian cancer to node width, and the total number of significantly deregulated studies in prostate cancer to its height. It is immediately evident which nodes have already been described to be up/down regulated in either one or both types of cancer. This can be useful to parallel the information already known from one cancer to the other. In addition, we can map the generated total of studies to node transparency, making genes with less disease evidence less obtrusive.

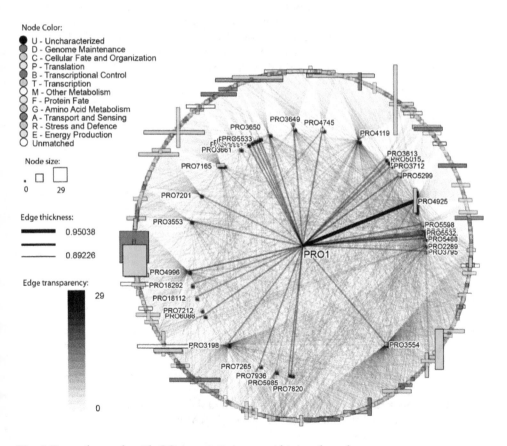

Fig. 6. Example graph with GO Annotation mapped to a color scheme.

We can also import structured data, in the form of GO attributes, retrieved from the I2D plug-in(Figure 6). We can view this data per individual node in the Node side panel, revealing the list of individual GO attributes and their descriptions. To get a graph-wide view of these attributes, we will add a filter to map the GO data to one of several categories, each with its own colour. The same result can be obtained by applying GO terms or other attributes, like pathways to which the node belongs, retrieved from other sources to the nodes as features and editing the filter in the desired way.

3.3 Importing drug-protein interactions

Finally, we will import a list of drugs and their gene targets as additional interactions. This expands our network to 2,707 nodes and 5,257 edges (Figure 7). Through a combination of manual layout and radial layout tools, we arrange the drugs in a circle around PRO1, its interactors, and their interactors from I2D. The edges connecting drugs to proteins are coloured blue to differentiate them from PPIs. To see the impact of individual drugs to this network, we map their degree to node size and transparency. Thus, large nodes represent drugs that target many of the proteins in the network. The top six of these drugs are labelled for convenience. Analogously, some proteins have a high degree of blue edges and connect to small nodes, such as ProX. These drugs show strong specificity to ProX. The initial data will be available in ASCII tab-delimited format and the final figure in NAViGaTOR 2 XML file at http://www.cs.utoronto.ca/~juris/data/intech12/.

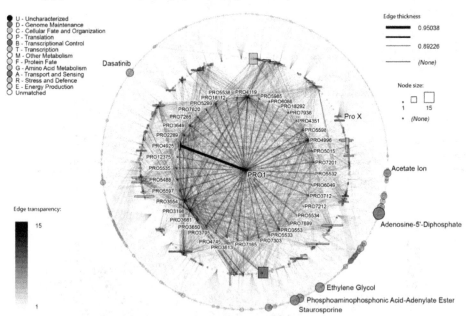

Fig. 7. Final graph, with drug interactions included and the size of nodes representing drugs derived from number of interactions within the graph. NAViGaTOR 2 XML file for the final figure is available in supplemental material
(http://www.cs.utoronto.ca/~juris/data/intech12/).

4. Conclusions

Integrated databases and resources are only useful when they can be effectively accessed, navigated and analyzed. Several biological network visualization tools are currently available, providing a diverse range of approaches and algorithms. While many existing visualization tools are effective and widely used, there are several critical areas where these applications require improvement. Scalability is essential to visualize the tens of thousands of known PPIs, which is a challenge for current layout algorithms and software. Biological graph drawing software must also be able to handle richly annotated data, including genomic and proteomic profiles, pathways, Gene Ontology annotations and data in PSI-MI and BioPAX formats, in addition to the vast quantity of microarray and proteomic data that is available.

Individual tools need a good balance of performance and useful features. The features that are needed for each use are highly dependent on the available data and the workflow. As in any creative activity, a tool may enable new workflows by providing novel features, but the tool may also lack certain important features, or offer features that are not needed. There is no single solution that satisfies all of these requirements at the present time, and as data and workflows change over time, network visualization tools must also evolve.

As the data grow more complex, the performance of layout algorithms will need to improve, and new options of differentiating multiple attributes will be required. As certain workflows become more main-stream, they may be turned into *analysis patterns* and implemented as plug-ins. Standardizing file formats, APIs and plug-ins will further intertwine existing tools, enabling their easier integration and specialization.

With new data and advances in computational biology, user tasks are modified, which must be reflected by types of algorithms that support analyses and the user interfaces that effectively enable them. New graph theory algorithms for faster and biologically meaningful network layouts and algorithms for network structure analysis will need to be integrated into network visualization tools. Importantly, none of these algorithms would make a broad difference unless a user interface appropriate for biologists is available (Viau et al., 2010).

5. Acknowledgment

This research was funded in part by Ontario Research Fund (GL2-01-030), Canada Foundation for Innovation (CFI #12301 and CFI #203383), and the Ontario Ministry of Health and Long Term Care. The views expressed do not necessarily reflect those of the OMOHLTC. CP was funded in part by Friuli Exchange Program. IJ is supported in part by the Canada Research Chair Program.

The authors would like to thank Max Kotlyar, Dan Strumpf, Fiona Broackes-Carter and the entire Jurisica lab for useful comments and discussions.

6. References

Alnemri, Emad, S., David J Livingston, Donald W Nicholson, Guy Salvesen, Nancy A Thornberry, Winnie W Wong, Junying Yuan (1986). Human ICE/CED-3 protease nomenclature, *Cell*, 87(2):171.

Aranda, B.,Blankenburg, H.,Kerrien, S.,Brinkman, F.S.L., Ceol, A., Chautard, E., Dana, J.M.,De Las Rivas, J., Dumousseau, M.,Galeota, E., Gaulton, A., Goll, J., Hancock, R.E.W., Isserlin, R., Jimenez, R.C., Kerssemakers, J., Khadake, J., Lynn, D.J., Michaut, M.,O'Kelly, G., Ono, K., Orchard, S., Prieto, C., Razick, S., Rigina, O., Salwinski, L., Simonovic, M., Velankar, S., Winter, A., Wu, G., Bader, G.D., Cesareni, G., Donaldson, I.M., Eisenberg, D., Kleywegt, G.J., Overington, J., Ricard-Blum, S., Tyers, M., Albrecht, M.,Hermjakob, H. (2011). PSICQUIC and PSISCORE: Accessing and scoring molecular interactions. *Nature Methods*, 8(7): 28-529.

Björn H. Junker, Christian Klukas and Falk Schreiber (2006). VANTED: A system for advanced data analysis and visualization in the context of biological networks. *BMC Bioinformatics*, 7:109

Brown, K.R., and Jurisica, I. (2005). Online Predicted Human Interaction Database. *Bioinformatics*, 21(9):2076-82.

Brown, K.R., and Jurisica, I. (2007). Unequal evolutionary conservation of human protein interactions in interologous networks. *Genome Biol*, 8(5):R95.

Brown, K.R., Otasek D, Ali M, McGuffin, M.J., Xie W, Devani B, van Toch I.L., and Jurisica I. (2009). NAViGaTOR: Network Analysis, Visualization and Graphing Toronto, *Bioinformatics*, 25(24): 3327-3329.

Demir E, Cary MP, Paley S, Fukuda K, Lemer C, Vastrik I, Wu G, D'Eustachio P, Schaefer C, Luciano J, Schacherer F, Martinez-Flores I, Hu Z, Jimenez-Jacinto V, Joshi-Tope G, Kandasamy K, Lopez-Fuentes AC, Mi H, Pichler E, Rodchenkov I, Splendiani A, Tkachev S, Zucker J, Gopinath G, Rajasimha H, Ramakrishnan R, Shah I, Syed M, Anwar N, Babur O, Blinov M, Brauner E, Corwin D, Donaldson S, Gibbons F, Goldberg R, Hornbeck P, Luna A, Murray-Rust P, Neumann E, Reubenacker O, Samwald M, van Iersel M, Wimalaratne S, Allen K, Braun B, Whirl-Carrillo M, Cheung KH, Dahlquist K, Finney A, Gillespie M, Glass E, Gong L, Haw R, Honig M, Hubaut O, Kane D, Krupa S, Kutmon M, Leonard J, Marks D, Merberg D, Petri V, Pico A, Ravenscroft D, Ren L, Shah N, Sunshine M, Tang R, Whaley R, Letovksy S, Buetow KH, Rzhetsky A, Schachter V, Sobral BS, Dogrusoz U, McWeeney S, Aladjem M, Birney E, Collado-Vides J, Goto S, Hucka M, Le Novère N, Maltsev N, Pandey A, Thomas P, Wingender E, Karp PD, Sander C, Bader GD. (2010). The BioPAX community standard for pathway data sharing. *Nature Biotechnology*. 28(9):935-42.

Djebbari, A., Ali, M., Otasek, D., Kotlyar. M., Fortney, K., Wong, S., Hrvojic, A. and Jurisica, I. (2011). NAViGaTOR: Scalable and Interactive Navigation and Analysis of Large Graphs. *Internet Mathematics*, 7(4):314-347.

Flicek P, Amode MR, Barrell D, Beal K, Brent S, Carvalho-Silva D, Clapham P, Coates G, Fairley S, Fitzgerald S, Gil L, Gordon L, Hendrix M, Hourlier T, Johnson N, Kähäri AK, Keefe D, Keenan S, Kinsella R, Komorowska M, Koscielny G, Kulesha E, Larsson P, Longden I, McLaren W, Muffato M, Overduin B, Pignatelli M, Pritchard B, Riat HS, Ritchie GR, Ruffier M, Schuster M, Sobral D, Tang YA, Taylor K, Trevanion S, Vandrovcova J, White S, Wilson M, Wilder SP, Aken BL, Birney E, Cunningham F, Dunham I, Durbin R, Fernández-Suarez XM, Harrow J, Herrero J, Hubbard TJ, Parker A, Proctor G, Spudich G, Vogel J, Yates A, Zadissa A, Searle SM. (2012). Ensembl. *Nucleic Acids Res*. [Epub ahead of print]

Gehlenborg N., O'Donoghue S.I., Baliga N.S., Goesmann A., Hibbs M.A., Kitano H., Kohlbacher O., Neuweger H., Schneider R., Tenenbaum D., Gavin A.C. (2003). Visualization of omics data for systems biology. *Nat Methods*, 7(3 Suppl):S56-68.

Helaers R, Bareke E, De Meulder B, Pierre M, Depiereux S, Habra N, Depiereux E. (2011). gViz, a novel tool for the visualization of co-expression networks. *BMC Res Notes*. 4(1):452.

Himsolt, M. (1996). GML: A portable Graph File Format. Syntax. Retrieved from http://www.fim.uni-passau.de/fileadmin/files/lehrstuhl/brandenburg/projekte/gml/gml-technical-report.pdf

Himsolt, M. (1996). GML: A portable Graph File Format. Syntax. Retrieved from http://www.fim.uni-passau.de/fileadmin/files/lehrstuhl/brandenburg/projekte/gml/gml-technical-report.pdf

Hu Z., Hung J.H., Wang Y., Chang Y.C., Huang C.L., Huyck M., DeLisi C. (2009). VisANT 3.5: multi-scale network visualization, analysis and inference based on the gene ontology. *Nucleic Acids Res*, 37, W115–W121.

Hu, Z., Mellor, J., Wu, J. and DeLisi, C. (2004). VisANT: an online visualization and analysis tool for biological interaction data. *BMC Bioinformatics*, 5, 17.

Jain E., Bairoch A., Duvaud S., Phan I., Redaschi N., Suzek B.E., Martin M.J., McGarvey P., Gasteiger E. (2009). Infrastructure for the life sciences: design and implementation of the UniProt website. *BMC Bioinformatics*, 10:136.

Junker B.H., Klukas C. & Schreiber F. (2006). VANTED: a system for advanced data analysis and visualization in the context of biological networks. *BMC Bioinformatics*, 7(109).

Kerrien S, Orchard S, Montecchi-Palazzi L, Aranda B, Quinn AF, Vinod N, Bader GD, Xenarios I, Wojcik J, Sherman D, Tyers M, Salama JJ , Moore S, Ceol A, Chatr-Aryamontri A, Oesterheld M, Stümpflen V, Salwinski L, Nerothin J, Cerami E, Cusick ME, Vidal M, Gilson M, Armstrong J, Woollard P, Hogue C, Eisenberg D, Cesareni G, Apweiler R, Hermjakob H (2007). Broadening the horizon--level 2.5 of the HUPO-PSI format for molecular interactions. *BMC Biol*. 5, 44

Kersey P. J., Duarte J., Williams A., Karavidopoulou Y., Birney E., Apweiler R. (2004). The International Protein Index: An integrated database for proteomics experiments. *Proteomics* 4(7): 1985-1988

Kreeger P.K., Lauffenburger D.A. (2010). Cancer systems biology: a network modeling perspective, *Carcinogenesis*, 31(1):2-8.

Longabaugh WJ. (2012). BioTapestry: a tool to visualize the dynamic properties of gene regulatory networks. *Methods Mol Biol*. 786:359-94.

Maglott D, Ostell J, Pruitt KD, Tatusova T. (2011). Entrez Gene: gene-centered information at NCBI. *Nucleic Acids Res*. 39(Database issue):D52-7.

McGuffin, M, and Jurisica, I. (2009). Interaction techniques for selecting and manipulating subgraphs in network visualizations. *IEEE Trans Vis Comput Graph*, 15 (6): 937-944.

Montojo J, Zuberi K, Rodriguez H, Kazi F, Wright G, Donaldson SL, Morris Q, Bader GD (2010). GeneMANIA Cytoscape plugin: fast gene function predictions on the desktop. *Bioinformatics*, 26: 22

Morrow JK, Tian L, Zhang S. (2010). Molecular networks in drug discovery. *Crit Rev Biomed Eng*. 38(2):143-56.

Niu, Y., Otasek, D., Jurisica, I. (2011). Evaluation of linguistic features useful in extraction of interactions from PubMed; Application to annotating known, high-throughput and predicted interactions in I2D. *Bioinformatics*, 26(1): 111-9.

Pavlopoulos G.A., Wegener A.L., Schneider R. (2008). A survey of visualization tools for biological network analysis, *BioData Min*, 1(12).

Remmerie N., De Vijlder T., Laukens K., Dang T.H., Lemière F., Mertens I., Valkenborg D., Blust R., Witters E. (2011). Next generation functional proteomics in non-model plants: A survey on techniques and applications for the analysis of protein complexes and post-translational modifications. *Phytochemistry*, 72(10):1192-218.

Safran M, Dalah I, Alexander J, Rosen N, Iny Stein T, Shmoish M, Nativ N, Bahir I, Doniger T, Krug H, Sirota-Madi A, Olender T, Golan Y, Stelzer G, Harel A and Lancet D. (2010). GeneCards Version 3: the human gene integrator *Database*; doi: 10.1093/database/baq020

Shannon P., Markiel A., Ozier O., Baliga N.S., Wang J.T., Ramage D., Amin N., Schwikowski B., Ideker T. (2003). Cytoscape: a software environment for integrated models of biomolecular interaction networks. *Genome Res*, 13:2498–2504.

Shirdel EA, Xie W, Mak TW, Jurisica I. (2011) NAViGaTing the micronome--using multiple microRNA prediction databases to identify signalling pathway-associated microRNAs. *PLoS One*. 6(2):e17429.

Smoot ME, Ono K, Ruscheinski J, Wang PL, Ideker T. (2011). Cytoscape 2.8: new features for data integration and network visualization. *Bioinformatics*. 1;27(3):431-2.

Stein A., Mosca R., Aloy P. (2011). Three-dimensional modeling of protein interactions and complexes is going 'omics. *Curr Opin Struct Biol*, 21(2):200-8.

Swainston N, Smallbone K, Mendes P, Kell D, Paton N. (2011). The SuBliMinaL Toolbox: automating steps in the reconstruction of metabolic networks. *J Integr Bioinform*. 8(2):186.

The Gene Ontology Consortium. (2000). Gene ontology: tool for the unification of biology. *Nat. Genet.* 25(1):25-9.

Viau, C., McGuffin, M J., Chiricota, Y., and Jurisica, I. (2010). The FlowVizMenu and parallel scatterplot matrix: Hybrid multidimensional visualizations for network exploration. *IEEE Trans Vis Comput Graph*, 16(6):1100-8.

Yu L.R. (2011). Pharmacoproteomics and toxicoproteomics: The field of dreams. *J Proteomics*, 74(12):2549-53.

8

A Survey on Evolutionary Analysis in PPI Networks

Pavol Jancura and Elena Marchiori
Radboud University Nijmegen
The Netherlands

1. Introduction

The analysis and application of the evolutionary information, as measured by means of the conservation of protein sequences, using protein-protein interaction (PPI) networks, has become one of the central research areas in systems biology from the last decade. It provides a promising approach for better understanding the evolution of living systems, for inferring relevant biological information about proteins, and for creating powerful protein interaction and function prediction tools. The aim of this survey is to give a general overview of the relevant literature and advances in the analysis and application of evolution in PPI networks. Due to the broad scope and vast literature on this subject, the present overview will focus on a representative selection of research directions and state-of-the-art methods to be used as a solid knowledge background for guiding the development of new hypothesis and methods aiming at the extraction and exploitation of evolutionary information in PPI networks.

This survey consists of two main parts (see Fig. 1). The first part deals with research works concerning the relation between evolution and the topological structures of a PPI network, in particular trying to discover and assess the evidence of such a relation and its strength at different granularity levels. Specifically, we consider works analysing evolution at the single protein level as well as at the level of a collection of proteins present in a PPI network. The second part of this survey describes works analysing how such evolutionary evidence can be exploited for knowledge discovery, in particular for inferring relevant biological information, such as protein interaction prediction and the discovery of functional modules conserved across multiple species.

The main terms and concepts underlying protein interaction and evolution which are used throughout the survey are summarized in the sequel. In general, a protein-protein interaction can represent different types of relations, such as a true physical bond or a functional interplay between proteins. Here, if not explicitly stated, a PPI represents a physical protein interaction as detected by experimental methods, such as yeast two-hybrid (Y2H) screening, co-immunoprecipitation or tandem affinity purification.

Two proteins are called *homologous* if they share high sequence similarity. There are two main types of homologous proteins: *orthologous* and *paralogous*. Here, for simplicity, we consider a protein pair to be orthologous if the proteins of the pair are from different species. We refer to the proteins of an orthologous pair as orthologs. Analogously, a protein pair is considered to be paralogous if its proteins belong to the same species, in this case their proteins are called paralogs. A general assumption is that the proteins of an orthologous pair originated from a common ancestor, having been separated in evolutionary time only by a speciation event, while paralogous proteins are the product of gene duplication without speciation. The concept of orthology can be directly extended to more than two species, where one can consider clusters of orthologous proteins containing at least one protein of each species.

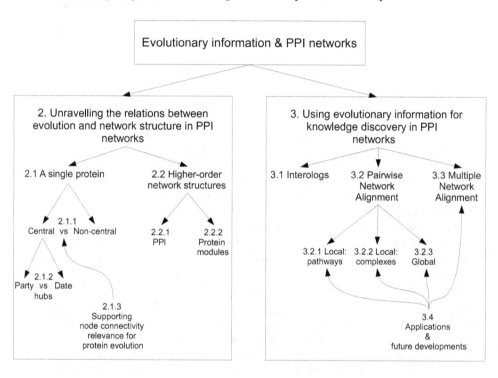

Fig. 1. The structure of the survey.

2. Unravelling the relations between evolution and network structure in PPI networks

We begin with a summary of those studies that involve the analysis of evolutionary information in a single PPI network. One can divide these works into the following two main groups. The first group studies evolutionary conservation with respect to topological properties of a PPI network. The second one primarily investigates the role of evolution with respect to the functional modules present in a PPI network.

The aim of the first group of studies is to describe how the topology of a single PPI network reflects the evolutionary signal present in the proteins it contains. This evolutionary signal is represented by the set of orthologs and it is retrieved with respect to a different species. Specifically, given a PPI network of the species to be investigated and a set of proteins of a distinct species, those proteins of the network being a part of orthologous pairs or clusters (resulting from a sequence comparison of proteins of the two or multiple species respectively) are considered to be source of the evolutionary or orthology signal in the network. Then, having established the orthology relationship between proteins of the two or multiple species, one can estimate the evolutionary rate or distance of aligned protein sequences (see e.g. Yang & Nielsen, 2000). The higher the rate, the faster is considered the evolution of proteins. Consequently, proteins which evolve slowly are well-conserved and a little or none change to them can be observed throughout the evolution. Other protein evolutionary measures have been also considered, as propensity for gene loss, evolutionary excess retention or protein age (see Table 1).

Type of evolutionary measure	Evolutionary measure	References
Evolutionary conservation	Evolutionary Rate	e.g. Yang & Nielsen (2000), Wall et al. (2005),
	Propensity for Gene Loss	Krylov et al. (2003)
	Evolutionary Excess Retention	Wuchty (2004)
	Phyletic Retention	Gustafson et al. (2006), Chen & Xu (2005), Fang et al. (2005)
Protein age classification	Time of Origin	Kunin et al. (2004)
	Protein Age Group	Ekman et al. (2006), Kim & Marcotte (2008)

Table 1. Measures of evolutionary signal at protein level

2.1 Relation between a single protein in a PPI network and evolution

Various features of a PPI network topology can be investigated with respect to evolutionary information; the first and simplest ones are measures acting on the single nodes of the network. One can associate with a node different topological measures which estimate the relative relevance of the node within the network, here called *centrality* or *connectivity* of a node.

A basic centrality measure of a node is its degree. The degree of a node is the number of edges containing the node or, in terms of a PPI network, it is the number of proteins with which the protein represented by the node in the network interacts. It has been observed that a protein degree distribution of PPI networks follows a power law and thus PPI networks fall into a class of scale-free networks (see e.g. Jeong et al., 2001). Scale-free networks have a few highly connected nodes, called hubs, and numerous less connected nodes, which mostly interact only with one or two nodes.

2.1.1 Essentiality, centrality and conservation of a protein

As a decade ago large protein physical interaction data were not yet available, researchers mainly focussed on the study of the correlation between importance of a protein function for a living cell (essentiality, dispensability) and its evolutionary conservation rate. The generally accepted premise is that essential genes or proteins should evolve at slower rates than non-essential ones (see e.g. Kimura, 1983). Although empirical studies have cast doubts on the validity of this hypothesis (see e.g. Hurst & Smith, 1999; Pal et al., 2003; Rocha & Danchin, 2004), in the end the vast majority and late evidences favour the existence of correlation between gene essentiality or dispensability and evolutionary conservation (see e.g. Fang et al., 2005; Fraser et al., 2002; 2003; Hahn & Kern, 2005; Hirsh & Fraser, 2001; 2003; Jordan et al., 2002; Krylov et al., 2003; Ulitsky & Shamir, 2007; Wall et al., 2005; Wang & Zhang, 2009; Waterhouse et al., 2011; Zhang & He, 2005). In particular, as recently stated by Wang & Zhang (2009), the correlation remains weak yet still conveniently sufficient for practical use.

After the growth of protein interaction data, also the correlation between essentiality and centrality, and evolutionary conservation and centrality started to be investigated. At first the *centrality-essentiality relationship* was mostly investigated by examining the degree of a node, proving the existence of the correlation (see e.g. Fraser et al., 2002; 2003; Hahn & Kern, 2005; Jeong et al., 2001; Krylov et al., 2003). However Coulomb et al. (2005) showed no correlation between essentiality and centrality, where centrality was assessed not only by the degree but also by higher order centrality measures, namely average neighbours' degree of a node and clustering coefficient of a node, suggesting that the correlation centrality-essentiality could be an artefact of the dataset. These findings were later supported by Gandhi et al. (2006) who considered a set of PPI networks and also did not observe any significant relationship between a node degree and the essentiality of the corresponding protein. Interestingly, Coulomb et al. (2005) did not test other centrality measures as betweenness and closeness, which showed a higher correlation with essentiality than just the simple degree (Hahn & Kern, 2005). Nevertheless, Batada, Hurst & Tyers (2006) reaffirmed the existence of the correlation between the node degree and essentiality taking into account Coulomb et al.'s concerns. However, Yu et al. (2008) again disputed the correlation using the compilation of Yeast high quality PPI data. Results contradicting this work appeared in two consecutive studies by Park & Kim (2009) and Pang, Sheng & Ma (2010). The first study (Park & Kim, 2009) considered also other centrality measures than just the degree of a node. As a result, the correlation could be successfully revealed, whereas the highest correlation was observed with measures based on betweenness and closeness, similarly to Hahn & Kern (2005). In the other study (Pang, Sheng & Ma, 2010) the newer, updated yeast PPI dataset was used and the correlation between degree of a node and its (protein) essentiality could be detected.

Although, the above works support that there is a connection between topological position of a node and functional importance, it seems one cannot explain this centrality-lethality rule just by the degree distribution (He & Zhang, 2006; Zotenko et al., 2008). This seems to be in accordance with the analysis conducted in (Lin et al., 2007) showing that protein domain complexity is not the single determinant of protein essentiality and that there is a correlation between the number of protein domains and the number of interactions (Schuster-Bockler & Bateman, 2007). In addition, Kafri et al. (2008) showed that highly connected essential proteins tend to have duplicates which can compensate their deletion thus decreasing the deleterious effect of their removal, a phenomenon that could possibly explain the findings that genes with no duplicates are more likely to be essential (Giaever et al., 2002). Therefore higher order topological features appear to be more appropriate for capturing gene essentiality, especially those based on node-betweenness and node-closeness (Hahn & Kern, 2005; Park & Kim, 2009; Yu et al., 2007), which are believed to estimate better the local connectivity or centrality of a node within the network. Moreover, these features also relate with gene expression (Krylov et al., 2003; Pang, Sheng & Ma, 2010; Yu et al., 2007).

We consider now works that analyse the correlation between evolution and centrality. Also in this case the two main features used to estimate this correlation are the degree of a node and the evolutionary rate. At first, it was hypothesized that proteins with a higher degree should evolve slower (Fraser et al., 2002). A main criticism to this hypothesis was based on the fact that the analysis conducted in (Fraser et al., 2002) did not take into account the presence of a possible bias and of noise in data obtained from high-throughput experiments (Bloom & Adami, 2003; Jordan et al., 2003a;b). Nevertheless Fraser et al. (2003), Fraser & Hirsh (2004) and Lemos et al. (2005) could confirm the existence of such correlation by taking into account these objections. Kim et al. (2007) also confirmed interconnection between centrality, essentiality and conservation and showed that peripheral proteins of the PPI network are under positive selection for species adaptation. Moreover, the link between the connectivity of a node and its evolutionary history was further substantiated by works studying the correlation between node degree and other evolutionary measures such as propensity for gene loss (Krylov et al., 2003), evolutionary excess retention (Wuchty, 2004) and protein age (Ekman et al., 2006; Kunin et al., 2004). However Batada, Hurst & Tyers (2006) again pointed to a lack of evidence for a significant correlation between the evolutionary rate and the connectivity of a node. Moreover, Makino & Gojobori (2006) classified proteins according to two criteria, clustering coefficient of a node and protein's multi-functionality, and showed that multi-functional proteins of sparse parts of yeast PPI network (with a low clustering coefficient) evolve at the slowest rate regardless of the degrees of the connectivity. This suggests that clustering coefficient is a better descriptor of protein evolution within the global network of protein interactions.

A possible explanation for these conflicting results was proposed by Saeed & Deane (2006) who showed that the strength and significance of the correlation between evolution and centrality varies depending upon the type of PPI data used. Also Saeed & Deane (2006) found that more accurate datasets demonstrate stronger correlations between connectivity and evolutionary rate than less accurate datasets. Another reason may be the existence of two distinct types of highly connected nodes, so-called *party* and *date hubs*, which appear to satisfy different evolutionary constraints.

2.1.2 Evolution of party and date hubs

Specifically, Han et al. (2004) observed a bimodal distribution of average Pearson correlation coefficients between the expression profiles of proteins and its interacting partners. This yielded a classification of hubs into party hubs, having similar co-expression profiles with their neighbours, and date hubs, having different co-expression profiles with their neighbours. As a consequence, party hubs tend to interact simultaneously ("permanently") with their partners and to connect proteins within functional modules while date hubs tend to interact with different partners at different time/space ("transiently") and to bridge different modules. Thus, one may also refer to party hubs as *intramodule* and to date hubs as *intermodule* (Fraser, 2005).

Fraser (2005) was the first to investigate the difference in evolution between date and party hubs and found that party hubs are highly evolutionary constrained, whereas date hubs are more evolutionary labile. This is clearly in accordance with findings of Mintseris & Weng (2005) who argued that residues in the interfaces of permanent protein interactions tend to evolve at a relatively slower rate, allowing them to co-evolve with their interacting partners, in contrast to the plasticity inherent in transient interactions, which leads to an increased rate of substitution for the interface residues and leaves little or no evidence of correlated mutations across the interface. The work of Fraser (2005) was, in addition, later corroborated by Bertin et al. (2007). Examining three dimensional properties of proteins also supported this hypothesis, as multi-interface hubs were found to be more evolutionary conserved and essential as well as more likely to correspond to party hubs (Kim et al., 2006). Defining singlish- and multi-Motif hubs further substantiated these findings, because multi-Motif hubs were found to be more evolutionary conserved, more essential and to correlate with multi-interface hubs (Aragues et al., 2007). In addition, other features as orderness of regions in protein sequences and the solvent accessibility of the amino acid residues was shown to be different between party and date hubs and to contribute in the lowering of the evolutionary rate of party hubs (Kahali et al., 2009). Recently, Mirzarezaee et al. (2010) applied feature selection methods and machine learning techniques to predict party and date hubs based on a set of different biological characteristics including amino acid sequences, domain contents, repeated domains, functional categories, biological processes, cellular compartments, etc.

However, other researchers disputed not only the evolutionary differences between party and date hubs but the existence of hub types as such (Agarwal et al., 2010; Batada, Reguly, Breitkreutz, Boucher, Breitkreutz, Hurst & Tyers, 2006; Batada et al., 2007). Indeed, some datasets do not exhibit clear or robust bimodal distribution of hubs' gene co-expression profiles (Agarwal et al., 2010) and in some cases there is even a complete lack of bimodality (Batada, Reguly, Breitkreutz, Boucher, Breitkreutz, Hurst & Tyers, 2006; Batada et al., 2007). Therefore, Pang, Cheng, Xuan, Sheng & Ma (2010) argue that the average Pearson correlation coefficient is a weak measure of whether a protein acts transiently or permanently with its interacting partners and they propose a new measure, a co-expressed protein-protein interaction degree. This measure estimates the actual number of partners with which a protein can permanently interact. One can interpret it as a degree of 'protein party-ness' and it offers more a continuum-like estimate of the protein's interaction property. This seems to be in accordance with Nooren & Thornton (2003) who suggest that rather a continuum range exists between distinct types of protein interactions and that their stability very much depends on the physiological conditions and environment.

Pang, Cheng, Xuan, Sheng & Ma (2010) firstly corroborated the results of Saeed & Deane (2006) on the correlation variations between connectivity and evolutionary rate of a protein on different datasets and then they showed that the co-expression-dependent node degree correlates significantly with the protein's evolutionary rate irrespectively of the specific dataset used. However, their topological measure is derived by using an external source of experimental data on gene expression. The further investigation on purely topological features of a PPI network which would distinguish transient and permanent interactions, and party and date hubs could bring more insights on how the evolutionary history of a protein is wired in its position within the network of all the protein interactions in an organism. In this perspective, network path-based measures, such as betweenness and closeness, seem to be promising (Yu et al., 2007). All the more, these measures also appear to relate to protein essentiality (Park & Kim, 2009; Yu et al., 2007) and it could clarify the link between essentiality and evolution as such. Thereafter, they could improve on the prediction of essential genes from the topology of a PPI network in combination with protein evolutionary information, such as phyletic retention (Gustafson et al., 2006), as already corroborated by several application of machine learning techniques for essential gene detection, prioritizing drug targets and determining virulence factors (see e.g. Chen & Xu, 2005; Deng et al., 2011; Doyle et al., 2010; Gustafson et al., 2006; McDermott et al., 2009).

2.1.3 Node connectivity is relevant for protein evolution

Since the factors relevant for protein evolution could be of a multiple character (Wolf et al., 2006), it is interesting to investigate whether protein connectivity plays a central or a more subtle role. In the latter case, the link between protein connectivity and evolution could be the results of spurious correlations due to other underlying biological processes (Bloom & Adami, 2003). In order to address this issue, the contribution of protein connectivity to protein evolutionary conservation has been also studied in an integrated way (Pal et al., 2006) using multidimensional methods such as principal component analysis (PCA) and principal component regression (PCR).

The first successful application of PCA was given by Wolf et al. (2006) on seven genome-related variables. The derived first component reflected a gene's 'importance' and confirmed positive correlation between lethality, expression levels and number of protein-protein interaction which at the same time constrained protein evolution measures. Interestingly, the component also showed that the number of paralogs positively contributes to gene essentiality, which contradicts the finding of Giaever et al. (2002) that non-duplicated genes tend to be essential. However, the study of Drummond et al. (2006) revealed by using PCR only single determinant of protein evolution, namely translational selection, which is almost entirely determined by the gene expression level, protein abundance, and codon bias. Later, Plotkin & Fraser (2007) re-examined the use of PCR method and showed noise in biological data can confound PCRs, leading to spurious conclusions. As a result, when they equalized for different amounts of noise across the predictor variables no single determinant of evolution could be found indicating that a variety of factors-including expression level, gene dispensability, and protein-protein interactions may independently affect evolutionary rates in yeast. This observation was further substantiated by a recent study (Theis et al., 2011) where 16 genomic variables were analysed using Bayesian PCA. The study supports the evidence for the three above-discussed correlations. It also demonstrates how different definitions of paralogs may lead to different conclusions on their effect on essentiality, and thus commenting on Wolf et al.'s conflicting result (Wolf et al., 2006).

2.2 Higher-order structures in a PPI network and evolution

Researchers have also focused on other topological structures of a PPI network than just a node and their relation to evolutionary conservation. With increasing topological complexity we may talk about a single protein-protein interaction (an edge in PPI network), topological motifs, and protein clusters or modules as detected by their interaction density or network traffic.

2.2.1 Evolution and protein-protein interaction

Unlike in the case of a single protein, where various well-established methods for measuring sequence evolution are developed, to the best of our knowledge only a recent attempt has been made in order to estimate the evolutionary rate of protein-protein interaction (Qian et al., 2011). However, this study is limited to a small set of PPIs in yeasts and can not be yet applied for large-scale studies due to the lack of data. Thus, the research has extensively focused on estimating correlated evolution of a protein pair and their functional or physical interaction (Pazos & Valencia, 2008).

It is generally assumed that proteins which co-evolve tend to participate together in a common biological function. This hypothesis is supported by many examples of functionally interacting protein families that co-evolve (see e.g. Galperin & Koonin, 2000; Moyle et al., 1994). Co-evolution of proteins may be assessed at sequence level (*sequence co-evolution*) by correlating evolutionary rates (Clark et al., 2011), or at gene family level (*gene family evolution*) by correlating occurrence vectors (Kensche et al., 2008). An occurrence vector or a phylogenetic profile (phyletic pattern) (Tatusov et al., 1997) is an encoding of protein's (homologue's) presence or absence within a given set of species of interest (Kensche et al., 2008). In general, the methods for correlating protein evolution have been successfully applied to predict a physical or functional interaction between proteins (Clark et al., 2011; Kensche et al., 2008), where sequence co-evolution is powerful in predicting the physical interaction and phylogenetic profiling is a good indicator of functional interplay between proteins in a broader sense. Large-scale co-evolutionary maps have also been constructed and analysed for better understanding the evolution of a species and its link to protein interactions (see e.g. Cordero et al., 2008; Tillier & Charlebois, 2009; Tuller et al., 2009). All these works suggest that the topology of PPIs should reflect the evolutionary processes behind the proteins which formed such network.

The first systematic study of linked genes and their evolutionary rates was done by Williams & Hurst (2000) who showed that the rates of linked genes are more similar than the rates of random pairs of genes. Pazos & Valencia (2001) performed the first successful large-scale prediction of physical PPIs based on sequence co-evolution by correlating phylogenetic trees. Another large-scale study by Kim et al. (2004) on domain structural data of interacting protein families also revealed their high co-evolution but also showed a high diversity in the correlation of rates of each family pair. Specifically, protein families with a greater number of domains were shown to be more likely to co-evolve. However, Hakes et al. (2007) argued that this correlation of evolutionary rates is not responsible for the covariation between functional residues of interacting proteins. Nevertheless, other studies have been able to predict interacting domains from co-evolving residues between domains or proteins (see e.g. Jothi et al., 2006; Yeang & Haussler, 2007) indicating that different organisms use the same 'building blocks' for PPIs and that the functionality of many domain pairs in mediating protein interactions is maintained in evolution (Itzhaki et al., 2006).

Another perspective on co-evolution of interacting partners was given by Mintseris & Weng (2005), who distinguished between transient and obligate interactions. The authors concluded that obligate complexes are likely to co-evolve with their interacting partners, while transient interactions with an increased evolutionary rate show only little evidence for a correlated evolution of the interacting interfaces. This observation was later corroborated by Brown

& Jurisica (2007) who analysed the presence of protein interactions across multiple species via orthology mapping and found that the greater the conservation of a protein interaction is, the higher the enrichment for stable complexes. Beltrao et al. (2009) also observed that stable interactions are more conserved than transient interactions, by studying evolution of interactions involved in phosphoregulation. Finally, Zinman et al. (2011) extracted protein modules from a yeast integrated protein interaction network using various source of PPI evidence, and showed that interactions within modules were much more likely to be conserved than interactions between proteins in different modules.

The preference of conserved protein interactions to be placed in modular parts of a network was also observed by Wuchty et al. (2006) by extending the paradigm of protein's connectivity and its evolutionary conservation to the connectivity of a protein-protein interaction. Specifically, they used the hypergeometric clustering coefficient to estimate the interaction cohesiveness of the PPI's neighbourhood and orthologous excess retention in order to asses the evolutionary conservation of PPIs. They used the same clustering coefficient as that given by the presence of orthologs of interacting proteins in another organism and showed that PPIs with highly clustered environment were accompanied by an elevated propensity for the corresponding proteins to be evolutionary conserved as well as preferably co-expressed (Wuchty et al., 2006). These findings are significant all the more they were shown to be stable under perturbations. This propensity of interacting proteins to be more conserved and prevalent among taxa was later confirmed by Tillier & Charlebois (2009) who used evolutionary distances to estimate the protein's conservation. Yet another perspective on conservation of PPIs was given by Kim & Marcotte (2008) who classified proteins into four groups (from oldest to youngest) according their age and found a unique interaction density pattern between different protein age groups, where the interaction density tends to be dense within the same group and sparse between different age groups.

2.2.2 Evolution and modularity of PPI networks

All the evidences above that PPIs whose proteins are evolutionary correlated tend to form stable complexes and to be embedded in cohesive areas of a network topology support the premise that modularity of PPI networks is maintained by evolutionary pressure (Vespignani, 2003). Indeed, when examining networks solely built from sequence co-evolution, gene context analysis or gene family evolution of completely sequenced genomes, one may observe that these networks exhibit high modularity with clusters corresponding to known functional modules, thus revealing the structure of cellular organization (Cordero et al., 2008; Tuller et al., 2009; von Mering et al., 2003).

Regarding the networks of physically interacting proteins, to the best of our knowledge the first direct evidence that evolution drives the modularity of PPI networks was provided by Wuchty et al. (2003). They looked beyond a single protein pair and studied the more complex patterns of interacting proteins, called topological motifs. In general, they found that, as the number of nodes in a motif and number of links among its constituents increase, a greater and stronger conservation of the proteins could be observed. This was corroborated by Vergassola et al. (2005) who focused on specific instances of motifs known as cliques. Cliques are topological patterns where all protein constituents interact with each other. Vergassola et al. (2005) provided evidence for co-operative co-evolution within cliques of interacting proteins. Later, Lee et al. (2006) investigated motifs at a higher resolution level, by defining for each motif different motif modes based on functional attributes of interacting proteins: again their findings indicated that motifs modes may very well represent the evolutionary conserved topological units of PPI networks. More recently, Liu et al. (2011) studied network motifs according to the age of their proteins and discovered that the proteins within motifs whose constituents are of the same age class tend to be densely interconnected, to co-evolve and to share the same biological functions. Moreover, these motifs tend to be within protein complexes.

The finding that modularity of PPI networks is constrained by evolution and that conserved interactions are enriched in dense motifs and regions of a PPI network also suggest that protein complexes present in such cohesive areas should be evolutionary driven (Jancura et al., 2012). As putative protein complexes can be extracted from a PPI network by means of clustering techniques, Jancura et al. (2012) detected such protein complexes in the PPI network consisting of only yeast proteins having an ortholog in another organism and compared them with those protein complexes derived either by using the global topology of a yeast PPI network or by using a network induced by randomly selected proteins. The in-depth examination of enriched functions in these three types of protein complexes revealed that evolutionary-driven complexes are functionally well differentiated from other two types of protein complexes found in the same interaction data. As a consequence, new complexes and protein function predictions could be unravelled from PPI data by using a standard clustering approach with the inclusion of evolutionary information. In addition, evolutionary-driven complexes were found to be differentially conserved, in particular some complexes were detected for all distinct set of orthologs as determined by comparison with different species, some exhibited only a subset of proteins identifiable in a complex across all species, and some complexes being found only for one specific set of orthologs. This suggests that presence of evolution in modularity of PPI networks is more versatile and flexible with different degrees of conservation.

The findings of Jancura et al. (2012) seem to conform with related studies that focused on evolutionary cohesiveness of protein functional modules in order to investigate whether a group of proteins which functionally interact, co-evolve more cohesively than a random group of proteins. Either known protein complexes and pathways were analysed (Fokkens & Snel, 2009; Seidl & Schultz, 2009; Snel & Huynen, 2004) or putative protein modules usually derived from integrated networks of functional link evidences (Campillos et al., 2006; Zhao et al., 2007; Zinman et al., 2011). A different strategy was employed by Yamada et al. (2006) who at first detected evolutionary modules which were afterwards compared with enzyme connectivity in a metabolic network.

Although the co-evolution of modules is assessed by the presence or absence of modules' constituents across a set of species, there is no standard method to measure the degree to which a module evolves cohesively (Fokkens & Snel, 2009). For instance, Snel & Huynen (2004) used the deviation of the number of modules' orthologs per species from the average number of modules' orthologs per species, whereas Campillos et al. (2006) measured the fraction of joined evolutionary events given the reconstructed, most parsimonious evolutionary scenario of the genes in a module over their phylogenetic profiles.

Despite this measures' diversity, the common conclusion is that the majority of modules evolve flexibly (Campillos et al., 2006; Fokkens & Snel, 2009; Seidl & Schultz, 2009; Snel & Huynen, 2004; Yamada et al., 2006). Also, it appears that curated modules evolve more cohesively than modules derived from high throughput interaction data (Fokkens & Snel, 2009; Seidl & Schultz, 2009; Snel & Huynen, 2004). Moreover, there is a different enrichment in functions which co-evolve. For example, biochemical pathways, certain metabolic and signalling processes, as well as core functions like transcription and translation, tend to have higher rate of evolutionary cohesiveness (Campillos et al., 2006; Fokkens & Snel, 2009; Zhao et al., 2007). This is also supported by methods which cluster phylogenetic profiles in order to detect biochemical pathways or to predict functional links and thus exploiting the predictive power of phylogenetic methods (Glazko & Mushegian, 2004; Li et al., 2009; Watanabe et al., 2008). These methods show a relatively good performance in characterizing biochemical pathways but seem to have a limited coverage for physically interacting proteins (Watanabe et al., 2008). A dubious result was reported on inter-connectivity of cohesive and flexible modules. Specifically, Fokkens & Snel (2009) demonstrated that components of cohesive modules are less likely to interact with each other than in the case of flexible modules, while two other studies (Campillos et al., 2006; Zinman et al., 2011) suggest cohesive modules to be more highly connected.

It is possible that the above studies underestimated the actual degree of evolutionary cohesiveness present in the modularity of protein interaction networks due to their conservative approach, the limitations in ortholog detection as well as the cohesiveness measures which are restricted to phylogenetic profiles. Nevertheless, they show that, as evolution is a complex process, its presence in modularity of protein interaction networks also exhibits a very complex nature, whose understanding is far from being complete. Evolution itself, indeed, can be expected to be asynchronous and heterotactous along the tree of life.

In general, the interim evidence shows different evolutionary pressure for different types of protein interactions and data. In particular, the slowly evolving interacting partners are enriched in stable, permanent complexes, and functional modules such as biochemical pathways and curated complexes exhibit higher evolutionary cohesiveness than high throughput complexes. It seems that the co-evolutionary degree of modules within PPI networks increases with greater integration of various sources of evidence for proteins to functionally interact (Zinman et al., 2011). Also, not all protein complexes and functional modules need to be co-evolutionary modules (Fokkens & Snel, 2009). There is a continuum from extremely conserved to rapidly changing modules, where those modules found to be co-evolving appear to be enriched in certain, specific functional categories (Campillos et al., 2006). In addition, the degree of conservation and co-evolution of functional modules within interaction networks seem to reflect cellular organization and their spatio-temporal characteristics. For instance, cohesive modules can be classified according to their evolutionary age as ancestral, intermediate and young, where one may observe ancient, ancestral modules to be highly conserved and perform essential, core processes such as information storage and metabolism of amino acids, while young modules are less conserved and responsible for the communication with the environment (Campillos et al., 2006). Therefore one might expect ancestral modules to contain static, obligate interactions as the proteins of essential functions tend to involve multiple domains with slow evolutionary rates, whereas young modules can be enriched with dynamic, transient interactions with less but fast evolving protein domains to allow adaptation to the environment.

3. Using evolutionary information for knowledge discovery in PPI networks

The tendency of functionally linked or physically interacting proteins and densely interacting motifs to exhibit correlated evolution and/or to be conserved across species is at the core of methods for inferring relevant biological information using PPI networks. Although such biological information can be limited and biased towards specific type of known interactions and protein functions, it allows one to infer new, unknown functions of proteins, to improve the understanding of biological systems, and to guide the discovery of drug-target interaction. In its basic form, the knowledge discovery process is based on the transfer of information involving a single interaction between two organisms, while in its most complex form it involves the identification and transfer of protein complexes across multiple species. In the sequel we summarize concepts and techniques used to achieve these goals, in particular the notions of "interologs" and of multiple PPI networks alignment.

3.1 Predicting protein interaction: Interologs

If two proteins physically interact in one species and they have orthologous counterparts in another species, it is likely that their orthologs interact in that species too. If such conserved interactions exist, they are called *interologs*. This simple method of protein interaction inference was firstly introduced and tested by Walhout et al. (2000) on proteins involved in vulval development of nematode worm, where potential interactions between these proteins were identified based on interactions of their orthologs in other species. Later, Matthews et al. (2001) performed a large-scale analysis of this inference technique using the yeast PPI network as a model and proteins of worm as a target. Although the success rate of detection of inferred interactions by Y2H analysis was between 16%-31%, it represented a 600-1100-fold increase compared to a conventional approach at that time (Matthews et al., 2001).

The interologs-based protein interaction prediction has become one of the standard methods for *in silico* PPI prediction. The method can be easily extended to more PPI data from multiple species. In particular, having two groups of orthologs, where each ortholog group contains proteins from the same N species, and observing an interaction between proteins of these orthologous groups in $(N-1)$ species, the interaction between proteins of the N-th species present in the ortholog groups can be predicted. This multidimensional character of interolog inference has been extensively used to predict and build databases of the whole interactome for various species, either as a stand alone approach or in combination with other *in silico* methods, which often integrate multiple data types including the gene co-expression, co-localization, functional category, the occurrence of orthologs and other genomic context methods. In this way researchers could provide, for instance, the first sketch of human interactome (Lehner & Fraser, 2004), build the interactome of plants (Geisler-Lee et al., 2007; Gu et al., 2011), and improve the understanding of processes in a malarial parasite (Pavithra et al., 2007) or in cancer (Jonsson & Bates, 2006). Also, three, up-to-date, tools have been recently implemented and made available to perform this inference task (Gallone et al., 2011; Michaut et al., 2008; Pedamallu & Posfai, 2010).

Several algorithmic enhancements of the interologs-based approach have been introduced since the first proposal of a systematic use of interolog inference (Matthews et al., 2001). For instance, Yu et al. (2004) have strengthen the definition of ortholog by using a reciprocal best-hit approach and compared it to the original one-way best-hit approach implemented by Matthews et al. (2001). In addition, they required a minimum level for a joint similarity of orthologous sequences in order to perform interolog mapping. Their method yielded a 54% accuracy in contrast to a 30% of the previous method by Matthews et al. (2001).

Other approaches exploited the knowledge on a higher conservation rate of PPIs in dense network motifs. For instance Huang et al. (2007) scored interologs according to the density of the topological pattern containing the respective PPI of the interolog in a model species as determined by the extraction of maximal quasi-cliques from the PPI network of the model species. This score was integrated with scores of other various features used for PPI prediction, such as tissue specificity, sub-cellular localization, interacting domains and cell-cycle stage. The use of multiple types of features was shown to yield more accurate predictions of PPIs in comparison with other interolog-based methods used to build interactome databases. More recently, Jaeger et al. (2010) proposed another interesting method based on two steps. First a set of all candidate interologs is built across the considered species. Next, interologs are assembled into maximal conserved and connected patterns by detecting frequent sub-graphs appearing in the interolog network of the candidate set. Only functionally coherent patterns were used for interolog inference.

The interolog concept was also modified and used in other ways and application domains. In particular, Tirosh & Barkai (2005) proposed a method to assess and increase the confidence of a predicted PPI by examining the co-expression of proteins of its potential interolog in other species. Chen et al. (2007) extended interolog mapping for homologous inference of interacting 3D-domains and they built a database of so-called 3D-interologs (Lo, Chen & Yang, 2010). Chen et al. (2009) used interologs to transfer conserved domain-domain interactions. Recently, Lo, Lin & Yang (2010) combined this interolog domain transfer with the former 3D-interolog detection technique and implemented an integrated tool for searching homologous protein complexes. Finally, Lee et al. (2008) exploited interologs to predict inter-species interactions.

Despite the successful use of interolog inference, a gap was observed between the actual, observed number of conserved interactions and the expected theoretical coverage (Gandhi et al., 2006; Lee et al., 2008). In order to test the reliability of interolog transfer, Mika & Rost (2006) performed a comprehensive validation of the method on several datasets. Their findings suggested that interolog transfers are only accurate at very high levels of sequence identity. In addition, they also compared the interolog transfer within species and across species. In the case of within-species interolog inference a PPI is transferred onto proteins which are sequence similar to the proteins of the considered PPI in the same species. Surprisingly, such paralogous interolog transfers of protein-protein interactions were shown to be significantly more reliable than the orthologous ones. This result was later substantiated by Saeed & Deane (2008), indicating that homology-based interaction prediction methods may yield better results when within-species interolog inference is also considered. In addition, Brown & Jurisica (2007) argued that one also needs to take into account whether all interactions have equal probability of being transferred between organisms. For example, the dynamic components of the interactomes are less likely to be accurately mapped from distantly related organisms. Moreover, there is apparent bias of interologs to be enriched in stable, permanent complexes (Brown & Jurisica, 2007), which is completely in accordance with findings on the different evolution of transient and permanent interactions. On the other hand, it is likely that the performance of interolog inference could be underestimated since its accuracy is assessed using experimentally tests based on Y2H techniques or high-throughput datasets with a high abundance in Y2H interactions, which were found to be highly enriched in transient and inter-complex connections (Yu et al., 2008).

3.2 Pairwise protein network alignment

Detection and transfer of an interolog between species have motivated the study and exploration of interspecies conservation of protein interactions on a global scale. In particular, instead of focusing on a conserved interaction alone one can compare and align whole interactome maps of distinct species, which mimics the idea behind sequence alignment methods. This approach gave a rise to so-called *network alignment* approach (Sharan & Ideker, 2006).

Using protein network alignment, one can either search for conserved functional network structures such as protein complexes and pathways, or identify functional orthologs across species. As a result this approach should provide a greater evidence and support for protein function and protein interaction prediction for yet uncharacterized or unknown biological processes. Protein network alignment methods can be classified into two main groups:*local network alignments* and *global network alignments*.

As most of the research attention has focused on comparing PPI networks of two different species, here we discuss the successive development of methods for, so-called, *pairwise network alignment*. In sequel we survey local pairwise alignments for detecting evolutionary conserved pathways, local pairwise alignments for detecting conserved protein complexes, and global pairwise network alignment techniques.

3.2.1 Local pairwise network alignment for pathway detection and query tasks

The main goal of local protein network alignment is to detect conserved pathways and protein complexes across species, by searching for local regions of input networks having both high topological similarity between the regions and high sequence similarity between proteins of these regions. The standard approach to this task consists of two main phases: *an alignment phase* and *a searching phase*. In the first phase a merged network representation of compared PPI networks is constructed, called *alignment or orthology graph*. The second phase performs a search for the structures of interest in the orthology graph. Each output result corresponds to a pair or multiplet of complexes or pathways which are evolutionary conserved across the two or more (PPI networks of the) species, respectively.

The first alignment method of whole PPI networks of two species using protein sequence similarity was introduced by Kelley et al. (2003). In this method, called *PathBLAST*, first a many-to-many mapping between proteins of the two species is determined by considering each pair of proteins with a sequence similarity higher than a given threshold as putative orthologs. Next, every orthologous pair is encoded in one alignment node of the new alignment graph and three types of edges (direct, gap and mismatch edge) are identified between these alignment nodes as follows. The direct edge corresponds to the case when a PPI between proteins of two orthologous pairs exists in the PPI networks of both species. The gap edge represents the case when in one species the respective proteins of alignments nodes are connected indirectly through a common neighbour. Finally, the mismatch edge between alignments nodes is formed if such indirect connection is found between the corresponding proteins in the PPI networks of both species. Gap and mismatch edges are used to describe possible evolutionary variations or account for experimental errors in data (Kelley et al., 2003). In the search phase, the alignment graph is turned into acyclic sub-graphs by random removal of alignment edges, which allows to extract high-scoring paths in linear time by a dynamic programming approach. The score of a path is computed as the sum of log probabilities of true orthology encoded in alignment nodes of the path and of true conserved interactions encoded by alignment edges contained in the path. Interestingly, the method was also applied to align a PPI network with its own copy. In this way they could identify conserved (paralogous) pathways within one species.

The work of Kelley et al. (2003) was followed by other alignment techniques for discovering conserved pathways based on evolutionary conservation. The main drawbacks of *PathBLAST* are that it detects conserved linear pathways in protein interaction data, which is represented as an undirected graph, and it has an exponentially worsening efficiency with the expected increasing length of a pathway to be detected. To circumvent these limitations Pinter et al. (2005) proposed an alignment technique designed explicitly for metabolic networks with directed links between enzymes. The method also handles more complex structures than a simple path, because the scoring of the alignment is based on sub-tree homeomorphism, which can be solved by an efficient deterministic approximation. Another enhancement for the pathway alignment problem was proposed by Wernicke & Rasche (2007) who designed a method that does not impose topological restrictions upon pathways and exploits the biological and local properties of pathways within the network. Another effective approach to metabolic network alignment was developed by Li et al. (2008) which uses an integrative score on compound and enzyme similarities. Pathway alignment has been further extensively investigated and various other techniques have been proposed (see e.g. Cheng et al., 2008; Koyutürk, Kim, Subramaniam, Szpankowski & Grama, 2006; Li et al., 2007).

The evolutionary mapping of *PathBLAST* can also be used to query a known pathway of one species into the PPI network of another species. However, due to limitations and algorithmic constraints of *PathBLAST*, many other methods have been developed with a focussed application of orthologous querying of biological functional complexes, and tools and web-services are available for querying general pathways and other types of protein functional modules across species (see e.g. Bruckner et al., 2009; Dost et al., 2008; Qian et al., 2009; Yang & Sze, 2007).

3.2.2 Local pairwise network alignment for protein complex detection

Another group of methods which followed *PathBLAST* focus on detection of conserved protein complexes across (PPI networks of two or more) species. As these methods compare networks of physical interactions, the identified complexes can be used for interolog prediction as well as for protein function prediction of yet uncharacterized proteins. The detected conserved complexes are either (putative) entire physical complexes or conserved parts of them.

To the best of our knowledge, the first method for detecting conserved complexes using pairwise comparison of PPI networks was introduced by Sharan, Ideker, Kelley, Shamir & Karp (2005) and called *NetworkBLAST*. It can be viewed as a direct extension of *PathBLAST* for the task of complex detection across species. The method employs a comprehensive probabilistic model for conservation of protein complexes and searches for heavy induced sub-graphs in the weighted orthology graph. As the maximal induced sub-graph problem is computationally intractable, *NetworkBLAST* employs a bottom-up greedy heuristic for this task.

Many alignment network techniques which followed *NetworkBLAST* are motivated by the computational intractability issue derived from the problem of a finding maximal common or induced sub-graph in an ortholog graph, and are based on different heuristics. For instance, Koyutürk, Kim, Topkara, Subramaniam, Grama & Szpankowski (2006) partitions the alignment graph into smaller clusters by performing an approximated balanced ratio-cut. In another method by Koyutürk, Kim, Subramaniam, Szpankowski & Grama (2006) the most frequent interaction motifs are extracted from an orthology-contracted graph. Liang et al. (2006) transforms the problem of maximal common sub-graph into the problem of finding all maximal cliques in the graph. Recently, Tian & Samatova (2009) introduced an algorithm based on detection of connected-components of the orthology graph solvable in a very efficient way.

Other researchers propose to restrict the search space to cope with intractability issue of searching phase instead of performing heavy heuristics. For example Li et al. (2007) pre-clusters one PPI network in order to detect candidate complexes which are afterwards aligned to the target species network with an exact integer programming algorithm. Jancura & Marchiori (2010) proposed a pre-processing algorithm based on detection of network hubs for dividing PPI networks, prior to their alignment, into smaller sub-networks containing potential conserved modules. Each possible pair of sub-networks can be later aligned with a state-of-the-art alignment method where the search phase can be performed by means of an exact algorithm, allowing one to perform network comparison in a fully modular fashion and possibly to parallelize the computation. An interesting modular approach was introduced by Narayanan & Karp (2007), where an orthology graph is not constructed but rather networks are compared and split consecutively in several recursive steps until all possible solutions, conserved sub-graphs, are found. Similarly, Gerke et al. (2007) only compares, but does not merge, local hub-centred regions of PPI networks as identified by clustering coefficients and node degrees. The method by Ali & Deane (2009) is again another example of approach where an alignment graph is not explicitly constructed; there interspecies protein similarities are considered as new edges in such a way that species PPI networks and similarity edges between them are encoded into a single global meta-graph which can be searched by standard clustering techniques.

There are also alignment methods which try to incorporate or use other types of information than just the one based on sequence similarity and interaction conservation. For instance, Guo & Hartemink (2009) exploited the findings on co-evolving interacting domains which mediate PPIs and, instead of using putatively homologous proteins for alignment, compares PPI networks across species according to conserved domains of protein-protein interactions. Ali & Deane (2009) propose a functionally guided alignment of PPI networks, where a scoring function incorporates not only sequence and topological similarity of aligned proteins but also their gene co-expression characteristics and coherence of functional annotations. Thus, the method can be seen as detecting functional modules shared across species rather than strictly evolutionary modules. Finally, Berg & Lässig (2006) developed a generalized alignment Bayesian method applicable to different biological networks.

Despite various pairwise alignment techniques have been introduced, only a few of them embody an evolutionary model of PPI networks in the scoring scheme of an alignment. Notably, Koyutürk, Kim, Topkara, Subramaniam, Grama & Szpankowski (2006) were the first to introduce a method that builds the orthology graph following the duplication/divergence model based on gene duplications. Another interesting method was proposed by Hirsh & Sharan (2007) who extended the probabilistic score of *NetworkBLAST* to asses the likelihood that two complexes originated from an ancestral complex in the common ancestor of the two species being compared under the evolutionary pressure of duplication and link dynamics events.

3.2.3 Global pairwise network alignment

In contrast to local network alignment, which uses many-to-many homologous mapping between proteins of distinct species to detect local conserved regions of a high topological similarity in the respective PPI networks, global protein network alignment uses this mapping to define an unique, globally optimal mapping across whole topologies of PPI networks (Singh et al., 2007), even if it were locally suboptimal in some regions of the networks. In the most strict form of this unique mapping each node in one input network is either matched to one node in the other input network or has no match in the other network. Thus the goal of global protein network alignment is to define functional orthologs across species, as the solution offers a way to resolve the ambiguity of orthology detection with the use of species interactome map. Naturally, as a by-product the global alignment can also identify conserved complexes or pathways.

To the best of our knowledge, the first method performing explicitly global alignment on pair of networks, called *IsoRank*, was introduced by Singh et al. (2007). Similarly to the local network alignment problem, the global network alignment problem is in general computationally intractable. As a consequence, *IsoRank* employs an approximation using an eigenvalue framework in a manner analogous to Google's PageRank algorithm.

Several advancements have naturally followed the introduction of *IsoRank*. For instance, Evans et al. (2008) proposed an asymmetric network matching algorithm based on a network simulation method called quantitative simulation, where a similarity score of a protein pair is iteratively updated by the similarity scores of their neighbours and vice versa until a unique global optimum is found. Other researchers focused more on formulating global alignment as combinatorial optimization problems. For instance Zaslavskiy et al. (2009) redefined the problem of global alignment as a standard graph matching problem and investigated methods using ideas and approaches from state-of-the-art graph matching techniques. Klau (2009) formalized global network alignment as an integer linear programming problem, where a near-optimal solution with a quality guarantee is found by solving a Lagrangian relaxation of the original optimization formulation. Recently, Chindelevitch et al. (2010) proposed a method where the global alignment is encoded as bipartite matching and applied a very efficient local optimization heuristic used for the well-known Travelling Salesman Problem.

3.3 Multiple protein network alignment

The methods on network alignment discussed so far perform alignment of two PPI networks of distinct species. The next natural extension is aligning more than two PPI networks, that is multiple network alignment. A first attempt to perform multiple local network alignment using three species was done by Sharan, Suthram, Kelley, Kuhn, McCuine, Uetz, Sittler, Karp & Ideker (2005), which exploited the scoring model of *NetworkBLAST*. However, the method scales exponentially with the number of input species and consequently it is ineffective for large scale comparisons.

Apart from the scalability problem, there are also other issues related to the problem of aligning more than two species. For instance, the putative orthologous mapping of certain proteins does not need to span across all species, meaning that proteins may be conserved only for a particular subset of species. This "orthology decay" is more evident when a large number of increasingly distant species are considered in the alignment. As a result, functional modules, such as pathways and complexes, can have a different degree of conservation, with some modules being strictly conserved across all species and some other modules being conserved only for a particular clade. Thus, a good alignment method should allow one to search for conserved modules at different degree of conservation. However, such requirement also increases the complexity of searching and consequently one may need to prune the number of all possible species combinations in alignment.

To the best of our knowledge, the first method capable of an efficient comparison of multiple PPI networks, called *Graemlin*, was introduced by Flannick et al. (2006). The alignment model of the method allows one to perform local as well as global alignment and is also applicable for querying tasks of particular biological modules of interest across PPI networks. It employs a rather involved scoring scheme which allows one to search for conserved pathways as well as for conserved complexes. It also outputs modules with a different conservation degree. *Graemlin* progressively aligns the closest pair of PPI networks according the species distance measured using a phylogenetic tree, until the last pair on the root of the tree is compared, corresponding to the most conserved parts of the aligned networks. The main disadvantage of this approach is that it involves to estimate many parameters. Recently, a supervised, automated parameter learner was proposed to lessen the burden of parameter tuning (Flannick et al., 2009).

Another phylogeny-guided local network alignment was proposed by Kalaev et al. (2008). Although the method uses the same probabilistic scoring for conserved complex as *NetworkBLAST*, it avoids its exponential scalability by redefining the alignment model such that it does not construct the merged representation of aligned networks but represents them as separate layers interconnected via orthologous mapping. Then a seed, that is, a group of putatively orthologous proteins spanning across all species, is selected using the species phylogeny and greedily expanded by adding other proteins being orthologous to each other in all respective species in order to maximize the alignment conservation score. The proposed method, however, identifies only protein complexes conserved across all species and does not detect complexes conserved only for a certain subset of species.

Notably, the functionally guided network alignment method of Ali & Deane (2009), previously mentioned as one of the methods for pairwise alignment, was also shown to perform efficiently local alignment of multiple networks.

All these multiple local network alignments do not reconstruct a plausible evolutionary history of PPI networks based on a model of evolution, although they might be phylogeny-aware. Motivated by this observation, Dutkowski & Tiuryn (2007) introduced a new multiple local network alignment method, called *CAPPI*, which from the given PPI networks of distinct species aims to reconstruct an ancient PPI network of the common ancestor. The method uses a Bayesian inference framework based on a duplication and divergence model of network evolution which mimics the processes by which most protein interactions are formed. After the reconstruction step, the ancestral network is decomposed into connected components which correspond to the ancestral modules of protein interactions and are projected back to the original networks to obtain the actual conserved network residues. Although the demonstrated application of the method was restricted to orthologous groups spanning across all species (Dutkowski & Tiuryn, 2007), to the best of our knowledge *CAPPI* is the only model-based approach for large-scale ancestral network reconstruction.

Among the multiple alignment methods above mentioned, only *Graemlin* was shown to perform a global multiple network alignment, yet it relies on a involved parameter estimation step and phylogeny-guided approximation. Recently Liao et al. (2009) developed another global alignment technique which is fully unsupervised and phylogeny-free. The method, called *IsoRankN*, is built on the *IsoRank* algorithm mentioned above (Singh et al., 2007) and its extension to the multiple global network alignment (Singh et al., 2008a). At first *IsoRankN* scores topological and sequence similarity matching between putatively orthologous proteins of each pair of input networks using *IsoRank*. Then, a maximum k-partite graph matching problem is formulated on the induced graph of pairwise alignment scores (Singh et al., 2008a) and the exact solution is approximated by a spectral graph partitioning algorithm. *IsoRankN* also effectively identifies one-to-one orthologous mappings for all subset of species and appears to out-perform *Graemlin* in terms of coverage and quality of functional enrichments.

3.4 Applications and future developments

Local and global alignment methods have been successfully applied to study evolution of species and to discover relevant biological knowledge. For example, Suthram et al. (2005) applied the network alignment of Sharan, Suthram, Kelley, Kuhn, McCuine, Uetz, Sittler, Karp & Ideker (2005) to examine the degree of conservation between the Plasmodium protein network and other model organisms, such as yeast, nematode worm, fruit fly and the bacterial pathogen Helicobacter pylori. They investigated whether the divergence of Plasmodium at the sequence level is reflected in the configuration of its protein network. Indeed, the alignments showed very little conservation suggesting that the patterns of protein interaction in Plasmodium, like its genome sequence, set it apart from other species (Suthram et al., 2005).

Another application of local network alignment was performed by Tan et al. (2007) who combined transcriptional regulatory interactions with protein-protein interactions and identified co-regulated complexes between yeast and fly revealing different conservation of their regulators. This finding advocates that PPI networks may evolve more slowly than transcriptional interaction networks. In addition, Schwartz et al. (2009) and Dutkowski & Tiuryn (2009) used conserved complexes detected by network alignments for protein interaction prediction in a manner similar to the interologs transfer approach and demonstrated their usefulness. In particular, Schwartz et al. (2009) provided a comprehensive experimental design which includes PPI prediction using network alignment, and demonstrated how effectively it reduces the cost of interactome mapping.

Furthermore, Bandyopadhyay et al. (2006) presented the first systematic identification of functional orthologs based on protein network comparison. They used the pairwise local alignment model of Kelley et al. (2003) to construct the orthology graph and then they resolved ambiguity of orthology mapping by fitting a logistic function previously trained on a known set of functional orthologs. In contrast, Singh et al. (2008b) predicted functional orthologs in unsupervised manner by using explicitly a global multiple network alignment method.

Finally, Kolar et al. (2008) performed a cross-species analysis of two herpes-viruses using the generalized Bayesian network alignment of Berg & Lässig (2006). Interestingly, the performed alignment employs in its probabilistic scoring system evolutionary rates of sequences and thus it goes beyond the narrow use of orthologous mapping as done in all other alignment techniques. The method predicted meaningful functional associations that could not be obtained from sequence or interaction data alone.

Despite the recent progress and increasing number of network alignment tools, their further development remains an ongoing research issue, in particular for multiple network comparison. Only a few methods perform the scoring of alignment according to evolutionary models and there is only one of them which fully reconstructs network evolutionary history. This clearly is in contrast with the numerous techniques for the reconstruction of evolutionary history of gene families. Also, actual alignment methods do not distinguish among diverse types of interactions, specifically between transient and permanent interactions. For example, the prior knowledge on different evolutionary behaviour of these types of physical interactions could be incorporated into a scoring scheme of alignment construction.

In addition, all but one network comparison methods just rely on the straightforward use of putative orthologous mapping as identified by sequence comparison or available in orthologous databases, but they do not employ evolutionary measures, such as evolutionary distances or retentions, which can be derived from the corresponding sequence alignments. These measures assess the level of evolutionary conservation and they could potentially improve the performance of network alignments.

Mostly all current applications of network alignments have worked with networks of physical interactome. However, the power of network alignment for functional annotation and other system biology applications could be explored when one performs comparison of more general, functional interaction networks. One may expect that such alignment could reveal a higher number of conserved modules as the interspecies conservation of modularity across protein networks increases with combined, integrated evidence for a pair of proteins to be functionally linked. Finally, all available methods here considered focused on conservation of modules but not on the more general concept of module evolutionary cohesiveness or co-evolution. The evolutionary cohesiveness can be assessed especially for the case of multiple alignments. Indeed, all conserved modules are inherently very cohesive, however not all evolutionary modules need to exhibit the correlated conservation at a level as expected by actual multiple network alignments. Protein functional modules differ in the degree of conservation and also in the degree of cohesiveness.

4. References

Agarwal, S., Deane, C. M., Porter, M. A. & Jones, N. S. (2010). Revisiting date and party hubs: Novel approaches to role assignment in protein interaction networks, *PLoS Comput Biol* 6(6): e1000817.

Ali, W. & Deane, C. M. (2009). Functionally guided alignment of protein interaction networks for module detection, *Bioinformatics* 25(23): 3166–3173.

Aragues, R., Sali, A., Bonet, J., Marti-Renom, M. A. & Oliva, B. (2007). Characterization of protein hubs by inferring interacting motifs from protein interactions, *PLoS Comput Biol* 3(9): e178.

Bandyopadhyay, S., Sharan, R. & Ideker, T. (2006). Systematic identification of functional orthologs based on protein network comparison, *Genome Research* 16(3): 428–435.

Batada, N. N., Hurst, L. D. & Tyers, M. (2006). Evolutionary and physiological importance of hub proteins, *PLoS Comput Biol* 2(7): e88.

Batada, N. N., Reguly, T., Breitkreutz, A., Boucher, L., Breitkreutz, B.-J., Hurst, L. D. & Tyers, M. (2006). Stratus not altocumulus: A new view of the yeast protein interaction network, *PLoS Biol* 4(10): e317.

Batada, N. N., Reguly, T., Breitkreutz, A., Boucher, L., Breitkreutz, B.-J., Hurst, L. D. & Tyers, M. (2007). Still stratus not altocumulus: Further evidence against the date/party hub distinction, *PLoS Biol* 5(6): e154.

Beltrao, P., Trinidad, J. C., Fiedler, D., Roguev, A., Lim, W. A., Shokat, K. M., Burlingame, A. L. & Krogan, N. J. (2009). Evolution of phosphoregulation: Comparison of phosphorylation patterns across yeast species, *PLoS Biol* 7(6): e1000134.

Berg, J. & Lässig, M. (2006). Cross-species analysis of biological networks by Bayesian alignment, *Proceedings of the National Academy of Sciences* 103(29): 10967–10972.

Bertin, N., Simonis, N., Dupuy, D., Cusick, M. E., Han, J.-D. J., Fraser, H. B., Roth, F. P. & Vidal, M. (2007). Confirmation of organized modularity in the yeast interactome, *PLoS Biol* 5(6): e153.

Bloom, J. & Adami, C. (2003). Apparent dependence of protein evolutionary rate on number of interactions is linked to biases in protein-protein interactions data sets, *BMC Evolutionary Biology* 3(1): 21.

Brown, K. & Jurisica, I. (2007). Unequal evolutionary conservation of human protein interactions in interologous networks, *Genome Biology* 8(5): R95.

Bruckner, S., Hüffner, F., Karp, R. M., Shamir, R. & Sharan, R. (2009). Torque: topology-free querying of protein interaction networks., *Nucleic Acids Research* 37(Web Server issue): W106–108.

Campillos, M., von Mering, C., Jensen, L. J. & Bork, P. (2006). Identification and analysis of evolutionarily cohesive functional modules in protein networks, *Genome Research* 16(3): 374–382.

Chen, C.-C., Lin, C.-Y., Lo, Y.-S. & Yang, J.-M. (2009). Ppisearch: a web server for searching homologous protein-protein interactions across multiple species, *Nucleic Acids Research* 37(suppl 2): W369–W375.

Chen, Y.-C., Lo, Y.-S., Hsu, W.-C. & Yang, J.-M. (2007). 3d-partner: a web server to infer interacting partners and binding models, *Nucleic Acids Research* 35(suppl 2): W561–W567.

Chen, Y. & Xu, D. (2005). Understanding protein dispensability through machine-learning analysis of high-throughput data, *Bioinformatics* 21(5): 575–581.

Cheng, Q., Berman, P., Harrison, R. & Zelikovsky, A. (2008). Fast alignments of metabolic networks, *BIBM '08: Proceedings of the 2008 IEEE International Conference on Bioinformatics and Biomedicine*, IEEE Computer Society, Washington, DC, USA, pp. 147–152.

Chindelevitch, L., Liao, C.-S. & Berger, B. (2010). Local optimization for global alignment of protein interaction networks, *Pacific Symposium on Biocomputing* 15: 123–132.

Clark, G. W., Dar, V.-u.-N., Bezginov, A., Yang, J. M., Charlebois, R. L. & Tillier, E. R. M. (2011). Using coevolution to predict protein-protein interactions, *in* G. Cagney, A. Emili & J. M. Walker (eds), *Network Biology*, Vol. 781 of *Methods in Molecular Biology*, Humana Press, pp. 237–256.

Cordero, O. X., Snel, B. & Hogeweg, P. (2008). Coevolution of gene families in prokaryotes, *Genome Research* 18(3): 462–468.

Coulomb, S., Bauer, M., Bernard, D. & Marsolier-Kergoat, M.-C. (2005). Gene essentiality and the topology of protein interaction networks, *Proceedings of the Royal Society B: Biological Sciences* 272(1573): 1721–1725.

Deng, J., Deng, L., Su, S., Zhang, M., Lin, X., Wei, L., Minai, A. A., Hassett, D. J. & Lu, L. J. (2011). Investigating the predictability of essential genes across distantly related organisms using an integrative approach, *Nucleic Acids Research* 39(3): 795–807.

Dost, B., Shlomi, T., Gupta, N., Ruppin, E., Bafna, V. & Sharan, R. (2008). Qnet: A tool for querying protein interaction networks, *Journal of Computational Biology* 15(7): 913–925.

Doyle, M., Gasser, R., Woodcroft, B., Hall, R. & Ralph, S. (2010). Drug target prediction and prioritization: using orthology to predict essentiality in parasite genomes, *BMC Genomics* 11(1): 222.

Drummond, D. A., Raval, A. & Wilke, C. O. (2006). A single determinant dominates the rate of yeast protein evolution, *Molecular Biology and Evolution* 23(2): 327–337.

Dutkowski, J. & Tiuryn, J. (2007). Identification of functional modules from conserved ancestral protein-protein interactions, *Bioinformatics* 23(13): i149–158.

Dutkowski, J. & Tiuryn, J. (2009). Phylogeny-guided interaction mapping in seven eukaryotes, *BMC Bioinformatics* 10(1): 393.

Ekman, D., Light, S., Björklund, A. K. & Elofsson, A. (2006). What properties characterize the hub proteins of the protein-protein interaction network of saccharomyces cerevisiae?, *Genome Biology* 7(6): R45.

Evans, P., Sandler, T. & Ungar, L. (2008). Protein-protein interaction network alignment by quantitative simulation, *BIBM '08: Proceedings of the 2008 IEEE International Conference on Bioinformatics and Biomedicine*, IEEE Computer Society, Washington, DC, USA, pp. 325–328.

Fang, G., Rocha, E. & Danchin, A. (2005). How essential are nonessential genes?, *Molecular Biology and Evolution* 22(11): 2147–2156.

Flannick, J., Novak, A., Do, C. B., Srinivasan, B. S. & Batzoglou, S. (2009). Automatic parameter learning for multiple local network alignment, *Journal of Computational Biology* 16(8): 1001–1022.

Flannick, J., Novak, A., Srinivasan, B. S., McAdams, H. H. & Batzoglou, S. (2006). Graemlin: General and robust alignment of multiple large interaction networks, *Genome Res.* 16(9): 1169–1181.

Fokkens, L. & Snel, B. (2009). Cohesive versus flexible evolution of functional modules in eukaryotes, *PLoS Comput Biol* 5(1): e1000276.

Fraser, H. B. (2005). Modularity and evolutionary constraint on proteins, *Nat Genet* 37(4): 351 – 352.

Fraser, H. B., Hirsh, A. E., Steinmetz, L. M., Scharfe, C. & Feldman, M. W. (2002). Evolutionary rate in the protein interaction network, *Science* 296(5568): 750–752.

Fraser, H. & Hirsh, A. (2004). Evolutionary rate depends on number of protein-protein interactions independently of gene expression level, *BMC Evolutionary Biology* 4(1): 13.

Fraser, H., Wall, D. & Hirsh, A. (2003). A simple dependence between protein evolution rate and the number of protein-protein interactions, *BMC Evolutionary Biology* 3(1): 11.

Gallone, G., Simpson, T. I., Armstrong, J. D. & Jarman, A. (2011). Bio::homology::interologwalk - a perl module to build putative protein-protein interaction networks through interolog mapping, *BMC Bioinformatics* 12(1): 289.

Galperin, M. Y. & Koonin, E. V. (2000). Who's your neighbor? new computational approaches for functional genomics, *Nat Biotech* 18(6): 609–613.

Gandhi, T. K. B., Zhong, J., Mathivanan, S., Karthick, L., Chandrika, K. N., Mohan, S. S., Sharma, S., Pinkert, S., Nagaraju, S., Periaswamy, B., Mishra, G., Nandakumar, K., Shen, B., Deshpande, N., Nayak, R., Sarker, M., Boeke, J. D., Parmigiani, G., Schultz, J., Bader, J. S. & Pandey, A. (2006). Analysis of the human protein interactome and comparison with yeast, worm and fly interaction datasets, *Nat Genet* 38(3): 285 – 293.

Geisler-Lee, J., O'Toole, N., Ammar, R., Provart, N. J., Millar, A. H. & Geisler, M. (2007). A predicted interactome for arabidopsis, *Plant Physiology* 145(2): 317–329.

Gerke, M., Bornberg-Bauer, E., Jiang, X. & Fuellen, G. (2007). Finding common protein interaction patterns across organisms, *Evolutionary bioinformatics online* 2: 45–52.

Giaever, G., Chu, A. M., Ni, L., Connelly, C., Riles, L., Veronneau, S., Dow, S., Lucau-Danila, A., Anderson, K., Andre, B., Arkin, A. P., Astromoff, A., El Bakkoury, M., Bangham, R., Benito, R., Brachat, S., Campanaro, S., Curtiss, M., Davis, K., Deutschbauer, A., Entian, K.-D., Flaherty, P., Foury, F., Garfinkel, D. J., Gerstein, M., Gotte, D., Guldener, U., Hegemann, J. H., Hempel, S., Herman, Z., Jaramillo, D. F., Kelly, D. E., Kelly, S. L., Kotter, P., LaBonte, D., Lamb, D. C., Lan, N., Liang, H., Liao, H., Liu, L., Luo, C., Lussier, M., Mao, R., Menard, P., Ooi, S. L., Revuelta, J. L., Roberts, C. J., Rose, M., Ross-Macdonald, P., Scherens, B., Schimmack, G., Shafer, B., Shoemaker, D. D., Sookhai-Mahadeo, S., Storms, R. K., Strathern, J. N., Valle, G., Voet, M., Volckaert, G., Wang, C.-y., Ward, T. R., Wilhelmy, J., Winzeler, E. A., Yang, Y., Yen, G., Youngman, E., Yu, K., Bussey, H., Boeke, J. D., Snyder, M., Philippsen, P., Davis, R. W. & Johnston, M. (2002). Functional profiling of the saccharomyces cerevisiae genome, *Nature* 418: 387–391.

Glazko, G. & Mushegian, A. (2004). Detection of evolutionarily stable fragments of cellular pathways by hierarchical clustering of phyletic patterns, *Genome Biology* 5(5): R32.

Gu, H., Zhu, P., Jiao, Y., Meng, Y. & Chen, M. (2011). Prin: a predicted rice interactome network, *BMC Bioinformatics* 12(1): 161.

Guo, X. & Hartemink, A. J. (2009). Domain-oriented edge-based alignment of protein interaction networks, *Bioinformatics* 25(12): i240–1246.

Gustafson, A., Snitkin, E., Parker, S., DeLisi, C. & Kasif, S. (2006). Towards the identification of essential genes using targeted genome sequencing and comparative analysis, *BMC Genomics* 7(1): 265.

Hahn, M. W. & Kern, A. D. (2005). Comparative genomics of centrality and essentiality in three eukaryotic protein-interaction networks, *Molecular Biology and Evolution* 22(4): 803–806.

Hakes, L., Lovell, S. C., Oliver, S. G. & Robertson, D. L. (2007). Specificity in protein interactions and its relationship with sequence diversity and coevolution, *Proceedings of the National Academy of Sciences* 104(19): 7999–8004.

Han, J.-D. J., Bertin, N., Hao, T., Goldberg, D. S., Berriz, G. F., Zhang, L. V., Dupuy, D., Walhout, A. J. M., Cusick, M. E., Roth, F. P. & Vidal, M. (2004). Evidence for dynamically organized modularity in the yeast protein-protein interaction network, *Nature* 430: 88–93.

He, X. & Zhang, J. (2006). Why do hubs tend to be essential in protein networks?, *PLoS Genet* 2(6): e88.

Hirsh, A. E. & Fraser, H. B. (2001). Protein dispensability and rate of evolution, *Nature* 411: 1046–1049.

Hirsh, A. E. & Fraser, H. B. (2003). Genomic function (communication arising): Rate of evolution and gene dispensability, *Nature* 421(6922): 497–498.

Hirsh, E. & Sharan, R. (2007). Identification of conserved protein complexes based on a model of protein network evolution, *Bioinformatics* 23(2): e170–176.

Huang, T.-W., Lin, C.-Y. & Kao, C.-Y. (2007). Reconstruction of human protein interolog network using evolutionary conserved network, *BMC Bioinformatics* 8(1): 152.

Hurst, L. D. & Smith, N. G. (1999). Do essential genes evolve slowly?, *Current biology* 9: 747–750.

Itzhaki, Z., Akiva, E., Altuvia, Y. & Margalit, H. (2006). Evolutionary conservation of domain-domain interactions, *Genome Biology* 7(12): R125.

Jaeger, S., Sers, C. & Leser, U. (2010). Combining modularity, conservation, and interactions of proteins significantly increases precision and coverage of protein function prediction, *BMC Genomics* 11(1): 717.

Jancura, P. & Marchiori, E. (2010). Dividing protein interaction networks for modular network comparative analysis, *Pattern Recognition Letters* 31(14): 2083 – 2096.

Jancura, P., Mavridou, E., Carrillo-De Santa Pau, E. & Marchiori, E. (2012). A methodology for detecting the orthology signal in a ppi network at a functional complex level, *BMC Bioinformatics* 13(Suppl 1). In press.

Jeong, H., Mason, S. P., Barabasi, A.-L. & Oltvai, Z. N. (2001). Lethality and centrality in protein networks, *Nature* 411: 41–42.

Jonsson, P. F. & Bates, P. A. (2006). Global topological features of cancer proteins in the human interactome, *Bioinformatics* 22(18): 2291–2297.

Jordan, I. K., Rogozin, I. B., Wolf, Y. I. & Koonin, E. V. (2002). Essential genes are more evolutionarily conserved than are nonessential genes in bacteria, *Genome Research* 12(6): 962–968.

Jordan, I. K., Wolf, Y. & Koonin, E. (2003a). Correction: No simple dependence between protein evolution rate and the number of protein-protein interactions: only the most prolific interactors tend to evolve slowly, *BMC Evolutionary Biology* 3(1): 5.

Jordan, I. K., Wolf, Y. & Koonin, E. (2003b). No simple dependence between protein evolution rate and the number of protein-protein interactions: only the most prolific interactors tend to evolve slowly, *BMC Evolutionary Biology* 3(1): 1.

Jothi, R., Cherukuri, P. F., Tasneem, A. & Przytycka, T. M. (2006). Co-evolutionary analysis of domains in interacting proteins reveals insights into domain-domain interactions mediating protein-protein interactions, *Journal of Molecular Biology* 362(4): 861 – 875.

Kafri, R., Dahan, O., Levy, J. & Pilpel, Y. (2008). Preferential protection of protein interaction network hubs in yeast: Evolved functionality of genetic redundancy, *Proceedings of the National Academy of Sciences* 105(4): 1243–1248.

Kahali, B., Ahmad, S. & Ghosh, T. C. (2009). Exploring the evolutionary rate differences of party hub and date hub proteins in saccharomyces cerevisiae protein-protein interaction network, *Gene* 429(1-2): 18 – 22.

Kalaev, M., Bafna, V. & Sharan, R. (2008). Fast and accurate alignment of multiple protein networks, *Research in Computational Molecular Biology*, pp. 246–256.

Kelley, B. P., Sharan, R., Karp, R. M., Sittler, T., Root, D. E., Stockwell, B. R. & Ideker, T. (2003). Conserved pathways within bacteria and yeast as revealed by global protein network alignment, *Proceedings of the National Academy of Science* 100: 11394–11399.

Kensche, P. R., van Noort, V., Dutilh, B. E. & Huynen, M. A. (2008). Practical and theoretical advances in predicting the function of a protein by its phylogenetic distribution, *Journal of The Royal Society Interface* 5(19): 151–170.

Kim, P. M., Korbel, J. O. & Gerstein, M. B. (2007). Positive selection at the protein network periphery: Evaluation in terms of structural constraints and cellular context, *Proceedings of the National Academy of Sciences* 104(51): 20274–20279.

Kim, P. M., Lu, L. J., Xia, Y. & Gerstein, M. B. (2006). Relating three-dimensional structures to protein networks provides evolutionary insights, *Science* 314(5807): 1938–1941.

Kim, W. K., Bolser, D. M. & Park, J. H. (2004). Large-scale co-evolution analysis of protein structural interlogues using the global protein structural interactome map (psimap), *Bioinformatics* 20(7): 1138–1150.

Kim, W. K. & Marcotte, E. M. (2008). Age-dependent evolution of the yeast protein interaction network suggests a limited role of gene duplication and divergence, *PLoS Comput Biol* 4(11): e1000232.

Kimura, M. (1983). *The Neutral Theory of Molecular Evolution*, Cambridge University Press.

Klau, G. (2009). A new graph-based method for pairwise global network alignment, *BMC Bioinformatics* 10(Suppl 1): S59.

Kolar, M., Lassig, M. & Berg, J. (2008). From protein interactions to functional annotation: graph alignment in herpes, *BMC Systems Biology* 2(1): 90.

Koyutürk, M., Kim, Y., Subramaniam, S., Szpankowski, W. & Grama, A. (2006). Detecting conserved interaction patterns in biological networks, *Journal of Computational Biology* 13(7): 1299–1322.

Koyutürk, M., Kim, Y., Topkara, U., Subramaniam, S., Grama, A. & Szpankowski, W. (2006). Pairwise alignment of protein interaction networks, *Journal of Compyutional Biology* 13(2): 182–199.

Krylov, D. M., Wolf, Y. I., Rogozin, I. B. & Koonin, E. V. (2003). Gene loss, protein sequence divergence, gene dispensability, expression level, and interactivity are correlated in eukaryotic evolution, *Genome Research* 13(10): 2229–2235.

Kunin, V., Pereira-Leal, J. B. & Ouzounis, C. A. (2004). Functional evolution of the yeast protein interaction network, *Molecular Biology and Evolution* 21(7): 1171–1176.

Lee, S.-A., Chan, C.-h., Tsai, C.-H., Lai, J.-M., Wang, F.-S., Kao, C.-Y. & Huang, C.-Y. (2008). Ortholog-based protein-protein interaction prediction and its application to inter-species interactions, *BMC Bioinformatics* 9(Suppl 12): S11.

Lee, W.-P., Jeng, B.-C., Pai, T.-W., Tsai, C.-P., Yu, C.-Y. & Tzou, W.-S. (2006). Differential evolutionary conservation of motif modes in the yeast protein interaction network, *BMC Genomics* 7(1): 89.

Lehner, B. & Fraser, A. (2004). A first-draft human protein-interaction map, *Genome Biology* 5(9): R63.

Lemos, B., Bettencourt, B. R., Meiklejohn, C. D. & Hartl, D. L. (2005). Evolution of proteins and gene expression levels are coupled in drosophila and are independently associated with mrna abundance, protein length, and number of protein-protein interactions, *Molecular Biology and Evolution* 22(5): 1345–1354.

Li, H., Kristensen, D. M., Coleman, M. K. & Mushegian, A. (2009). Detection of biochemical pathways by probabilistic matching of phyletic vectors, *PLoS ONE* 4(4): e5326.

Li, Y., de Ridder, D., de Groot, M. & Reinders, M. (2008). Metabolic pathway alignment between species using a comprehensive and flexible similarity measure, *BMC Systems Biology* 2(1): 111.

Li, Z., Zhang, S., Wang, Y., Zhang, X.-S. & Chen, L. (2007). Alignment of molecular networks by integer quadratic programming, *Bioinformatics* 23(13): 1631–1639.

Liang, Z., Xu, M., Teng, M. & Niu, L. (2006). Comparison of protein interaction networks reveals species conservation and divergence, *BMC Bioinformatics* 7(1): 457.

Liao, C.-S., Lu, K., Baym, M., Singh, R. & Berger, B. (2009). IsoRankN: spectral methods for global alignment of multiple protein networks, *Bioinformatics* 25(12): i253–258.

Lin, Y.-S., Hwang, J.-K. & Li, W.-H. (2007). Protein complexity, gene duplicability and gene dispensability in the yeast genome, *Gene* 387(1-2): 109 – 117.

Liu, Z., Liu, Q., Sun, H., Hou, L., Guo, H., Zhu, Y., Li, D. & He, F. (2011). Evidence for the additions of clustered interacting nodes during the evolution of protein interaction networks from network motifs, *BMC Evolutionary Biology* 11(1): 133.

Lo, Y.-S., Chen, Y.-C. & Yang, J.-M. (2010). 3d-interologs: an evolution database of physical protein- protein interactions across multiple genomes, *BMC Genomics* 11(Suppl 3): S7.

Lo, Y.-S., Lin, C.-Y. & Yang, J.-M. (2010). Pcfamily: a web server for searching homologous protein complexes, *Nucleic Acids Research* 38(suppl 2): W516–W522.

Makino, T. & Gojobori, T. (2006). The evolutionary rate of a protein is influenced by features of the interacting partners, *Molecular Biology and Evolution* 23(4): 784–789.

Matthews, L. R., Vaglio, P., Reboul, J., Ge, H., Davis, B. P., Garrels, J., Vincent, S. & Vidal, M. (2001). Identification of potential interaction networks using sequence-based searches for conserved protein-protein interactions or "interologs", *Genome Research* 11(12): 2120–2126.

McDermott, J. E., Taylor, R. C., Yoon, H. & Heffron, F. (2009). Bottlenecks and hubs in inferred networks are important for virulence in salmonella typhimurium, *Journal of Computational Biology* 16: 169–180.

Michaut, M., Kerrien, S., Montecchi-Palazzi, L., Chauvat, F., Cassier-Chauvat, C., Aude, J.-C., Legrain, P. & Hermjakob, H. (2008). Interoporc: automated inference of highly conserved protein interaction networks, *Bioinformatics* 24(14): 1625–1631.

Mika, S. & Rost, B. (2006). Protein-protein interactions more conserved within species than across species, *PLoS Comput Biol* 2(7): e79.

Mintseris, J. & Weng, Z. (2005). Structure, function, and evolution of transient and obligate protein-protein interactions, *Proceedings of the National Academy of Sciences* 102(31): 10930–10935.

Mirzarezaee, M., Araabi, B. & Sadeghi, M. (2010). Features analysis for identification of date and party hubs in protein interaction network of saccharomyces cerevisiae, *BMC Systems Biology* 4(1): 172.

Moyle, W. R., Campbell, R. K., Myers, R. V., Bernard, M. P., Han, Y. & Wang, X. (1994). Co-evolution of ligand-receptor pairs, *Nature* 368(6468): 251–255.

Narayanan, M. & Karp, R. M. (2007). Comparing protein interaction networks via a graph match-and-split algorithm, *Journal of Computational Biology* 14(7): 892–907.

Nooren, I. M. & Thornton, J. M. (2003). Diversity of protein-protein interactions, *EMBO J* 22(14): 3486–3492.

Pal, C., Papp, B. & Hurst, L. D. (2003). Genomic function (communication arising): Rate of evolution and gene dispensability, *Nature* 421(6922): 496–497.

Pal, C., Papp, B. & Lercher, M. J. (2006). An integrated view of protein evolution, *Nat Rev Genet* 7: 337–348.

Pang, K., Cheng, C., Xuan, Z., Sheng, H. & Ma, X. (2010). Understanding protein evolutionary rate by integrating gene co-expression with protein interactions, *BMC Systems Biology* 4(1): 179.

Pang, K., Sheng, H. & Ma, X. (2010). Understanding gene essentiality by finely characterizing hubs in the yeast protein interaction network, *Biochemical and Biophysical Research Communications* 401(1): 112 – 116.

Park, K. & Kim, D. (2009). Localized network centrality and essentiality in the yeast-protein interaction network, *PROTEOMICS* 9(22): 5143–5154.

Pavithra, S. R., Kumar, R. & Tatu, U. (2007). Systems analysis of chaperone networks in the malarial parasite plasmodium falciparum, *PLoS Comput Biol* 3(9): e168.

Pazos, F. & Valencia, A. (2001). Similarity of phylogenetic trees as indicator of protein-protein interaction, *Protein Engineering* 14(9): 609–614.

Pazos, F. & Valencia, A. (2008). Protein co-evolution, co-adaptation and interactions, *EMBO J* 27(20): 2648–2655.

Pedamallu, C. S. & Posfai, J. (2010). Open source tool for prediction of genome wide protein-protein interaction network based on ortholog information, *Source Code for Biology and Medicine* 5(1): 8.

Pinter, R. Y., Rokhlenko, O., Yeger-Lotem, E. & Ziv-Ukelson, M. (2005). Alignment of metabolic pathways, *Bioinformatics* 21(16): 3401–3408.

Plotkin, J. B. & Fraser, H. B. (2007). Assessing the determinants of evolutionary rates in the presence of noise, *Molecular Biology and Evolution* 24(5): 1113–1121.

Qian, W., He, X., Chan, E., Xu, H. & Zhang, J. (2011). Measuring the evolutionary rate of protein-protein interaction, *Proceedings of the National Academy of Sciences* 108(21): 8725–8730.

Qian, X., Sze, S.-H. & Yoon, B.-J. (2009). Querying Pathways in Protein Interaction Networks Based on Hidden Markov Models, *Journal of Computational Biology* 16(2): 145–157.

Rocha, E. P. C. & Danchin, A. (2004). An analysis of determinants of amino acids substitution rates in bacterial proteins, *Molecular Biology and Evolution* 21(1): 108–116.

Saeed, R. & Deane, C. (2006). Protein protein interactions, evolutionary rate, abundance and age, *BMC Bioinformatics* 7(1): 128.

Saeed, R. & Deane, C. (2008). An assessment of the uses of homologous interactions, *Bioinformatics* 24(5): 689–695.

Schuster-Bockler, B. & Bateman, A. (2007). Reuse of structural domain-domain interactions in protein networks, *BMC Bioinformatics* 8(1): 259.

Schwartz, A. S., Yu, J., Gardenour, K. R., Finley Jr, R. L. & Ideker, T. (2009). Cost-effective strategies for completing the interactome, *Nat Meth* 6(1): 55–61.

Seidl, M. & Schultz, J. (2009). Evolutionary flexibility of protein complexes, *BMC Evolutionary Biology* 9(1): 155.

Sharan, R. & Ideker, T. (2006). Modeling cellular machinery through biological network comparison, *Nature Biotechnology* 24(4): 427–433.

Sharan, R., Ideker, T., Kelley, B. P., Shamir, R. & Karp, R. M. (2005). Identification of protein complexes by comparative analysis of yeast and bacterial protein interaction data, *Journal of Computational Biology* 12(6): 835–846.

Sharan, R., Suthram, S., Kelley, R. M., Kuhn, T., McCuine, S., Uetz, P., Sittler, T., Karp, R. M. & Ideker, T. (2005). From the Cover: Conserved patterns of protein interaction in multiple species, *Proceedings of the National Academy of Sciences* 102(6): 1974–1979.

Singh, R., Xu, J. & Berger, B. (2007). Pairwise global alignment of protein interaction networks by matching neighborhood topology, *Research in Computational Molecular Biology* pp. 16–31.

Singh, R., Xu, J. & Berger, B. (2008a). Global alignment of multiple protein interaction networks, *Pacific Symposium on Biocomputing* 13: 303–314.

Singh, R., Xu, J. & Berger, B. (2008b). Global alignment of multiple protein interaction networks with application to functional orthology detection, *Proceedings of the National Academy of Sciences* 105(35): 12763–12768.

Snel, B. & Huynen, M. A. (2004). Quantifying modularity in the evolution of biomolecular systems, *Genome Research* 14(3): 391–397.

Suthram, S., Sittler, T. & Ideker, T. (2005). The plasmodium protein network diverges from those of other eukaryotes, *Nature* 438(7064): 108–112.

Tan, K., Shlomi, T., Feizi, H., Ideker, T. & Sharan, R. (2007). Transcriptional regulation of protein complexes within and across species, *Proceedings of the National Academy of Sciences* 104(4): 1283–1288.

Tatusov, R. L., Koonin, E. V. & Lipman, D. J. (1997). A genomic perspective on protein families, *Science* 278(5338): 631–637.

Theis, F. J., Latif, N., Wong, P. & Frishman, D. (2011). Complex principal component and correlation structure of 16 yeast genomic variables, *Molecular Biology and Evolution* 28(9): 2501–2512.

Tian, W. & Samatova, N. F. (2009). Pairwise alignment of interaction networks by fast identification of maximal conserved patterns, *Pacific Symposium on Biocomputing* 14: 99–110.

Tillier, E. R. & Charlebois, R. L. (2009). The human protein coevolution network, *Genome Research* 19(10): 1861–1871.

Tirosh, I. & Barkai, N. (2005). Computational verification of protein-protein interactions by orthologous co-expression, *BMC Bioinformatics* 6(1): 40.

Tuller, T., Kupiec, M. & Ruppin, E. (2009). Co-evolutionary networks of genes and cellular processes across fungal species, *Genome Biology* 10(5): R48.

Ulitsky, I. & Shamir, R. (2007). Pathway redundancy and protein essentiality revealed in the saccharomyces cerevisiae interaction networks, *Mol Syst Biol* 3: 1–7.

Vergassola, M., Vespignani, A. & Dujon, B. (2005). Cooperative evolution in protein complexes of yeast from comparative analyses of its interaction network, *PROTEOMICS* 5(12): 3116–3119.

Vespignani, A. (2003). Evolution thinks modular, *Nature Genetics* 35(2): 118–119.

von Mering, C., Zdobnov, E. M., Tsoka, S., Ciccarelli, F. D., Pereira-Leal, J. B., Ouzounis, C. A. & Bork, P. (2003). Genome evolution reveals biochemical networks and functional modules, *Proceedings of the National Academy of Sciences* 100(26): 15428–15433.

Walhout, A. J. M., Sordella, R., Lu, X., Hartley, J. L., Temple, G. F., Brasch, M. A., Thierry-Mieg, N. & Vidal, M. (2000). Protein interaction mapping in c. elegans using proteins involved in vulval development, *Science* 287(5450): 116–122.

Wall, D. P., Hirsh, A. E., Fraser, H. B., Kumm, J., Giaever, G., Eisen, M. B. & Feldman, M. W. (2005). Functional genomic analysis of the rates of protein evolution, *Proceedings of the National Academy of Sciences* 102(15): 5483–5488.

Wang, Z. & Zhang, J. (2009). Why is the correlation between gene importance and gene evolutionary rate so weak?, *PLoS Genet* 5(1): e1000329.

Watanabe, R., Morett, E. & Vallejo, E. (2008). Inferring modules of functionally interacting proteins using the bond energy algorithm, *BMC Bioinformatics* 9(1): 285.

Waterhouse, R. M., Zdobnov, E. M. & Kriventseva, E. V. (2011). Correlating traits of gene retention, sequence divergence, duplicability and essentiality in vertebrates, arthropods, and fungi, *Genome Biology and Evolution* 3: 75–86.

Wernicke, S. & Rasche, F. (2007). Simple and fast alignment of metabolic pathways by exploiting local diversity, *Bioinformatics* 23(15): 1978–1985.

Williams, E. J. B. & Hurst, L. D. (2000). The proteins of linked genes evolve at similar rates, *Nature* 407(6806): 900–903.

Wolf, Y. I., Carmel, L. & Koonin, E. V. (2006). Unifying measures of gene function and evolution, *Proceedings of the Royal Society B: Biological Sciences* 273(1593): 1507–1515.

Wuchty, S. (2004). Evolution and topology in the yeast protein interaction network, *Genome Research* 14(7): 1310–1314.

Wuchty, S., Barabasi, A.-L. & Ferdig, M. (2006). Stable evolutionary signal in a yeast protein interaction network, *BMC Evolutionary Biology* 6(1): 8.

Wuchty, S., Oltvai, Z. N. & Barabási, A.-L. (2003). Evolutionary conservation of motif constituents in the yeast protein interaction network, *Nature Genetics* 35(2): 176–179.

Yamada, T., Kanehisa, M. & Goto, S. (2006). Extraction of phylogenetic network modules from the metabolic network, *BMC Bioinformatics* 7(1): 130.

Yang, Q. & Sze, S.-H. (2007). Path matching and graph matching in biological networks, *Journal of Computational Biology* 14(1): 56–67.

Yang, Z. & Nielsen, R. (2000). Estimating synonymous and nonsynonymous substitution rates under realistic evolutionary models, *Molecular Biology and Evolution* 17(1): 32–43.

Yeang, C.-H. & Haussler, D. (2007). Detecting coevolution in and among protein domains, *PLoS Comput Biol* 3(11): e211.

Yu, H., Braun, P., Yildirim, M. A., Lemmens, I., Venkatesan, K., Sahalie, J., Hirozane-Kishikawa, T., Gebreab, F., Li, N., Simonis, N., Hao, T., Rual, J.-F., Dricot, A., Vazquez, A., Murray, R. R., Simon, C., Tardivo, L., Tam, S., Svrzikapa, N., Fan, C., de Smet, A.-S., Motyl, A., Hudson, M. E., Park, J., Xin, X., Cusick, M. E., Moore, T.,

Boone, C., Snyder, M., Roth, F. P., Barabási, A.-L., Tavernier, J., Hill, D. E. & Vidal, M. (2008). High-quality binary protein interaction map of the yeast interactome network, *Science* 322(5898): 104–110.

Yu, H., Kim, P. M., Sprecher, E., Trifonov, V. & Gerstein, M. (2007). The importance of bottlenecks in protein networks: Correlation with gene essentiality and expression dynamics, *PLoS Comput Biol* 3(4): e59.

Yu, H., Luscombe, N. M., Lu, H. X., Zhu, X., Xia, Y., Han, J.-D. J., Bertin, N., Chung, S., Vidal, M. & Gerstein, M. (2004). Annotation transfer between genomes: Protein-protein interologs and protein-dna regulogs, *Genome Research* 14(6): 1107–1118.

Zaslavskiy, M., Bach, F. & Vert, J.-P. (2009). Global alignment of protein-protein interaction networks by graph matching methods, *Bioinformatics* 25(12): i259–1267.

Zhang, J. & He, X. (2005). Significant impact of protein dispensability on the instantaneous rate of protein evolution, *Molecular Biology and Evolution* 22(4): 1147–1155.

Zhao, J., Ding, G.-H., Tao, L., Yu, H., Yu, Z.-H., Luo, J.-H., Cao, Z.-W. & Li, Y.-X. (2007). Modular co-evolution of metabolic networks, *BMC Bioinformatics* 8(1): 311.

Zinman, G., Zhong, S. & Bar-Joseph, Z. (2011). Biological interaction networks are conserved at the module level, *BMC Systems Biology* 5(1): 134.

Zotenko, E., Mestre, J., O'Leary, D. P. & Przytycka, T. M. (2008). Why do hubs in the yeast protein interaction network tend to be essential: Reexamining the connection between the network topology and essentiality, *PLoS Comput Biol* 4(8): e1000140.

Permissions

The contributors of this book come from diverse backgrounds, making this book a truly international effort. This book will bring forth new frontiers with its revolutionizing research information and detailed analysis of the nascent developments around the world.

We would like to thank Weibo Cai, PhD and Hao Hong, PhD, for lending their expertise to make the book truly unique. They have played a crucial role in the development of this book. Without their invaluable contribution this book wouldn't have been possible. They have made vital efforts to compile up to date information on the varied aspects of this subject to make this book a valuable addition to the collection of many professionals and students.

This book was conceptualized with the vision of imparting up-to-date information and advanced data in this field. To ensure the same, a matchless editorial board was set up. Every individual on the board went through rigorous rounds of assessment to prove their worth. After which they invested a large part of their time researching and compiling the most relevant data for our readers. Conferences and sessions were held from time to time between the editorial board and the contributing authors to present the data in the most comprehensible form. The editorial team has worked tirelessly to provide valuable and valid information to help people across the globe.

Every chapter published in this book has been scrutinized by our experts. Their significance has been extensively debated. The topics covered herein carry significant findings which will fuel the growth of the discipline. They may even be implemented as practical applications or may be referred to as a beginning point for another development. Chapters in this book were first published by InTech; hereby published with permission under the Creative Commons Attribution License or equivalent.

The editorial board has been involved in producing this book since its inception. They have spent rigorous hours researching and exploring the diverse topics which have resulted in the successful publishing of this book. They have passed on their knowledge of decades through this book. To expedite this challenging task, the publisher supported the team at every step. A small team of assistant editors was also appointed to further simplify the editing procedure and attain best results for the readers.

Our editorial team has been hand-picked from every corner of the world. Their multi-ethnicity adds dynamic inputs to the discussions which result in innovative outcomes. These outcomes are then further discussed with the researchers and contributors who give their valuable feedback and opinion regarding the same. The feedback is then collaborated with the researches and they are edited in a comprehensive manner to aid the understanding of the subject.

Apart from the editorial board, the designing team has also invested a significant amount of their time in understanding the subject and creating the most relevant covers. They scrutinized every image to scout for the most suitable representation of the subject and create an appropriate cover for the book.

The publishing team has been involved in this book since its early stages. They were actively engaged in every process, be it collecting the data, connecting with the contributors or procuring relevant information. The team has been an ardent support to the editorial, designing and production team. Their endless efforts to recruit the best for this project, has resulted in the accomplishment of this book. They are a veteran in the field of academics and their pool of knowledge is as vast as their experience in printing. Their expertise and guidance has proved useful at every step. Their uncompromising quality standards have made this book an exceptional effort. Their encouragement from time to time has been an inspiration for everyone.

The publisher and the editorial board hope that this book will prove to be a valuable piece of knowledge for researchers, students, practitioners and scholars across the globe.

List of Contributors

Hao Hong and Weibo Cai
Departments of Radiology and Medical Physics, University of Wisconsin - Madison, Madison, WI, USA

Shreya Goel
Centre of Nanotechnology, Indian Institute of Technology, Roorkee, India

Jifeng Zhang
Department of Analytical and Formulation Sciences, Amgen Inc., Thousand Oaks, California, USA

Poluri Maruthi Krishna Mohan
Department of Chemistry & Chemical Biology, Rutgers University, New Jersey, USA

José Campos-Terán
Departamento de Procesos y Tecnología, DCNI, Universidad Autónoma Metropolitana, Mexico

Paola Mendoza-Espinosa and Jaime Mas-Oliva
Instituto de Fisiología Celular, Universidad Nacional Autónoma de México, Mexico

Rolando Castillo
Instituto de Física, Universidad Nacional Autónoma de México, México, D.F., México

Takeshi Hase
Department of Bioinformatics, Graduate School of Biomedical Science, Tokyo Medical and Dental University, Tokyo,
The Systems Biology Institute, Tokyo,

Yoshihito Niimura
Department of Bioinformatics, Medical Research Institute, Tokyo Medical and Dental University, Tokyo, Japan

Jose Ramon Blas
Universidad de Castilla-La Mancha, Spain

Joan Segura
University of Leeds, UK

Narcis Fernandez-Fuentes
University of Leeds, UK
Aberystwyth University, UK

David Otasek
Ontario Cancer Institute the Campbell Family Institute for Cancer Research, University Health Network, Toronto, Ontario, Canada

Chiara Pastrello
CRO Aviano, National Cancer Institute, Aviano, Italy
Ontario Cancer Institute the Campbell Family Institute for Cancer Research, University Health Network, Toronto, Ontario, Canada

Igor Jurisica
Department of Computer Science and Medical Biophysics, University of Toronto, Toronto, Canada
Ontario Cancer Institute the Campbell Family Institute for Cancer Research, University Health Network, Toronto, Ontario, Canada

Pavol Jancura and Elena Marchiori
Radboud University Nijmegen, The Netherlands

Printed in the USA
CPSIA information can be obtained
at www.ICGtesting.com
JSHW011358221024
72173JS00003B/336